TURING'S WORLD 3.0

CSLI Lecture Notes
Number 35

TURING'S WORLD 3.0
for the Macintosh®
An Introduction to Computability Theory

JON BARWISE & JOHN ETCHEMENDY

CSLI PUBLICATIONS
Center for the Study of
Language and
Information

STANFORD, CALIFORNIA

CSLI was founded early in 1983 by researchers from Stanford University, SRI International, and Xerox PARC to further research and development of integrated theories of language, information, and computation. CSLI headquarters and the publication offices are located at the Stanford site.

CSLI/SRI International
333 Ravenswood Avenue
Menlo Park, CA 94025

CSLI/Stanford
Ventura Hall
Stanford, CA 94305

CSLI/Xerox PARC
3333 Coyote Hill Road
Palo Alto, CA 94304

Printed in the United States
01 00 99 98   5 4 3

Cover design by John Etchemendy

Library of Congress Cataloging-in-Publication Data

Barwise, Jon
    Turing's World 3.0 / Jon Barwise and John Etchemendy.
      p. cm. -- (CSLI lecture notes ; no. 35)
    Includes index.
    ISBN 1-881526-10-0 (pbk.)
    1. Computer science. 2. Logic, Symbolic and mathematical. 3. Turing machines. 4. Turing's world. I. Etchemendy, John, 1952– . II. Title. III. Series.
    QA76.B3115 1993
    511.3--dc20   93-3761
                                           CIP

Please send comments and suggestions to one or both of us at:

Professor Jon Barwise
Department of Computer Science
Lindley Hall
Indiana University
Bloomington, IN 47405

Professor John Etchemendy
CSLI
Ventura Hall
Stanford University
Stanford, CA 94305-4115

To Jason Strober and Rick Wong

Contents

How to Use This Book

Welcome to Turing's World, the world of computation. Computer courseware is designed to help students use computers to learn more dynamically, and so more quickly and efficiently. In that regard, Turing's World is no exception. What's different, though, is the subject matter. Rather than economics, mathematics, physics, or political science, the subject matter of Turing's World is computation itself.

Turing's World is a self-contained introduction to some of the fundamental notions of logic and computer science, the notions of a *Turing machine* and a *finite state machine*. For many years, the theory of Turing machines was studied without the help of real computers. We developed Turing's World to introduce people to this theory in a way that takes advantage of the power of personal computers—in particular, of the Apple Macintosh®. The computer allowed us to provide the user with a friendly environment to build, debug, and run Turing machines. Using Turing's World, people with no previous experience can produce very sophisticated Turing machines in a matter of hours. More important, they gain a grasp of the power of these simple machines in a way never before possible.

Turing's World was originally conceived and implemented at Stanford University for use as a supplement to an intermediate level logic course, "Computability and Logic," but the resulting program and book have been used in many different ways, both by individuals working on their own and in classroom settings at many different universities and high schools.

The success of the original program inspired us to improve and expand it. Several users wrote suggesting changes and enhancements. The current version of the program, Turing's World 3.0, incorporates most of these suggestions. The most obvious changes have to do with the ability to design finite state machines and nondeterministic machines. But just below the surface are a host of other changes that make the program easier and more fun to use. These include features like being able to copy and paste portions of a state diagram, being able to write directly on the tape from the keyboard, and a much more flexible text field capability, for use in annotating state diagrams. In addition, there is now an even faster "lightning" mode for running machines.

About the book

The book gives a brief explanation of Turing machines, and a smattering of the historical background that led to the concept. Mainly, though, it teaches you how to use the program to design and build Turing machines, and in so doing gives you a good sense of just what Turing machines are, how they work, and why the notion is so powerful. To increase the program's usefulness, we have included over a hundred exercises, from very simple to quite challenging. A user who did most of these exercises would have a first-rate grounding in computability theory, from which to pursue more advanced topics. Toward this end, we have included an annotated list of further readings at the end of Chapter 1.

Chapter 1 is devoted to explaining the basic idea of a Turing machine, and sketching why this notion has been so important in the fields of logic, mathematics, and computer science. If you are already familiar with this material, you can skip this chapter and get right to playing around with the program. If you skip the chapter but later find you can't understand why the sample machines in Chapter 2 run the way they do, you can always go back and read Chapter 1 later.

The best way to learn is by doing, so we've written the bulk of the book to involve the user actively, in the style of a "tutorial." In Chapter 2 we explain how to run some prepackaged Turing machines, machines we have included on the Turing's World disk. Then in Chapters 3, 4, and 5 we lead the user step by step through the construction of some Turing machines.

After reading Chapter 3, you will be able to build your own machines. The later chapters give you more powerful techniques for building and debugging machines, and for delving into finite state and nondeterministic machines. Chapter 8 contains additional exercises which gradually lead you through the construction of some very sophisticated Turing machines while introducing you to some of the important theoretical topics in the field of computation.

While the tutorial portion of the book is intended to be worked through from beginning to end, we've also tried to arrange it so that it is easy to find help with specific problems you might run up against. The main guides here are the Table of Contents and the Index. Look through them now, and whenever you need help with something that's slipped your mind. If you're having problems and don't know exactly where to look, try looking under "problems" in the index.

To the individual user

With Turing's World, anyone with a Macintosh can learn the rudiments of computability theory and have fun doing it. There is no need to use it in a classroom setting or in conjunction with any other text. We hope that you will have so much fun and become so fascinated with the subject that you will want to pursue in greater depth some of the topics presented here. The list of readings at the end of Chapter 1 will point you in some interesting directions.

In writing this book, we have integrated the instructions for using the program with introductory material explaining the basic notions of computability. Thus you should work through the book from the beginning, at least through Chapter 6. At that point, you could skip ahead to Chapter 8, which contains many interesting problems touching on fundamental topics within computability theory. Only two of the sections in Chapter 8, covering finite state and nondeterministic machines, require that you have worked through Chapter 7.

The problems in Chapter 8 are not presented in order of difficulty, except within individual sections. If you get bogged down in one section, move on to the early problems in the next section. Some of the problems are significant enough that we have called them **Projects**. A project might well take a day or more to complete, but they are certainly worth the effort.

Have fun.

To the classroom user

We use Turing's World as part of our courses (with homework being done and graded on the Macintosh), but it can also be used as an aid in designing Turing machines for courses where this software is not required. It can also be used by instructors who want to demonstate Turing machines at work, even if the student's do not have access to the program.

An excellent use of the program is to have students work together in small groups on the more challenging exercises, those identified as **Projects**. You can also give additional exercises from other books or that you make up yourself.

We have not given explicit instructions in the exercises about how students should name the solution files they create. You will want to think about how you are going to have students turn in their solutions, whether on disks or on a file server, and give them explicit instructions on how to identify and organize them.

Instructor's disk

An instructor's disk is available to teachers using this book in a class. The instructor's disk allows you to grade most of the Turing's World problems automatically, directly off the student's disks. To receive the disk, write to: CSLI Publications, Ventura Hall, Stanford University, Stanford, CA 94305-4115.

Acknowledgements

Our main debt goes to the programmers whose skill, originality, and hard work made Turing's World a reality. The original version of Turing's World was programmed by Rick Wong, Sha Xin Wei, and Atty Mullins under the supervision of Steve Loving, as part of the Stanford Faculty Author Development Program. *Turing's World 3.0* is a complete reworking of the original program, and was carried out by Jason Strober, largely under the direction of John Etchemendy. Financial support for the project came from the Center for the Study of Language and Information. We would like to thank all of our programmers, as well as Stanford University for providing the kind of atmosphere where creative ideas in education are encouraged and supported.

We are also indebted to Professor Kenneth Regan, who passed on to us a number of interesting problems and machines built in an earlier version of Turing's World, and to Greg Wheless for sending a Busy Beaver machine that he designed in a class taught by Professor Richard Grandy at Rice University.

We would be delighted to receive interesting problems and machines from our readers. Please send them to either of us at the addresses below.

Jon Barwise	John Etchemendy
Department of Computer Science	**CSLI**
Lindley Hall	Ventura Hall
Indiana University	Stanford University
Bloomington, IN 47405	Stanford, CA 94305

Turing's World on the World-Wide Web

Students and instructors can find up-to-date information about Turing's World on the World-Wide Web at the following URL:

http://www-csli.stanford.edu/hp/

This web site contains information about both the book and the program, and allows owners of the program to download the latest version. Connect to the web site and follow the link to "up-to-date version information."

1

About Turing Machines

Computability theory, as a branch of logic, came into existence before the electronic digital computers of which it is a theory. Indeed, the concept of a Turing machine was introduced by the English logician Alan Turing in 1936, and the results he proved about them were a major step in the development of modern day programmable computers. It is a classic case of pure research having enormous but unforeseen practical consequences.

1.1 What can't computers do?

Computers are everywhere these days, in virtually every store, office, bank, and school. They run our microwave ovens, control the brakes on our cars, and book reservations to concerts. Chances are they play chess better than you; they certainly play better than either of us. They are used in medical diagnosis, they control the flight of rockets into space and, for better or worse, play a key and ever increasing military role. The list goes on and on. Indeed, this very book was composed and typeset on a computer.

It is hard to imagine anyone reading this book who hasn't had frequent contact with computers. So you are probably aware of the great power of computers, and probably also aware of computer failures. No one can use a computer very long without something going wrong. That is what computability theory is all about:

the tremendous power of computation, and also its inherent limitations.

It isn't hard to be impressed by all that computers can do, even if they do break down on us from time to time. Overly impressed, perhaps. Nowadays too many people have the attitude that if something takes a lot of thought, or has to be done many times, then someone should just write a computer program to do it. In some domains this is fine, but computers do have their limitations. And because computers play such an important role in our lives, a role that will only increase in the years to come, it is vital for informed people to understand something of what computers can and can't do. Only if we understand both the power and the limitations of computation will we be free to use computers in safe, appropriate ways, free both from the huckster's exaggerated claims and the doomsayer's dire predictions about computers. Computability theory addresses one important facet of that understanding.

1.1.1 Three steps to a working program

Getting a computer to carry out some real-world task can be broken down into three parts. First, one has to turn the task into a mathematical problem. That is, one must build a mathematical model of the real-world task. Second, one must solve the mathematical problem in a very special way. One must find a solution that a patient, obedient child, one with lots of time and memory but no originality or knowledge of the world, could use. For that is basically what computers are. This sort of solution to a mathematical problem is called an *algorithm*. Algorithms are routine, step-by-step procedures for solving mathematical problems, procedures that can be carried out by children or perhaps even highly trained monkeys. But more to the point, they are procedures that can be executed by machines. Finally, once an algorithm has been found to solve the mathematical task, it must be implemented on a real computer to solve the original problem.

These three steps are crucial to any real-world computer application:

1. mathematically *modeling* the problem,
2. constructing an *algorithm*, and
3. *implementing* the algorithm on a real computer.

The people doing the work may not be conscious of the three separate steps, but the steps are always there, implicit or explicit. Designing any computer application in a responsible manner requires an understanding of each of these steps, what they amount to, how things can go wrong, and, finally, the inherent limitations each step imposes on our effective use of computers in the world.

1.1.2 Bugs

Notice that the three steps we've described are things that people have to do. And where people are involved, mistakes can be made. The blanket terms "computer error" and "bug" are used to describe all kinds of mistakes, from building a bad model, to making a mistake in the design of the algorithm, to implementing the algorithm incorrectly. Of course sometimes there are real *computer* errors, in the sense that something goes wrong with the actual machine, with the "hardware." In fact, the term "bug" comes from the days when insects sometimes got into the works, and people often blamed all their problems on such creatures. But most bugs, so-called "computer" errors, are really human failings, mistakes made in one or more of the steps we've outlined here.

One typical mistake is to use a mathematical model inadequate to the task at hand. For example, if you tried to model the number of dollars in bank accounts using natural (i.e. whole) numbers, you would miss all the cents. Of course, few of us would make so blatant a mistake as this, but mistakes that arise in modeling the problem are actually a very common source of computer bugs. Any model involves a great deal of idealization and simplification, and this is fine as long as the features ignored are irrelevant to the problem addressed, or nearly so. Unfortunately, it's often hard to predict which features can be safely ignored and which cannot, and as the problem we're addressing gets more complex, it gets harder and harder to spot potential inadequacies in our mathematical model.

Mistakes in the design of algorithms, unlike those in the modeling, are really mathematical errors: not errors of calculation, usually, but errors of proof. The designer may think that the algorithm devised solves the mathematical problem, while it really doesn't. He or she either didn't bother to prove that it solves the problem it was meant to, or the proof contained a mistake. In real life, sad to say, people seldom try to prove that an algorithm is correct. Usually, they think that it is obviously correct, that it obviously does what the designer had in mind, even when it doesn't.

Mistakes of implementation are usually called programming errors. Somewhere between the mathematical algorithm one finally settles on and the program one actually writes, something goes awry. Anyone who has done even the most rudimentary programming has encountered these sorts of errors. And they are hardly limited to inexperienced programmers. Nowadays, many computer programs are enormous things, much longer than *War and Peace* or *The Lord of the Rings*. It is not at all surprising that they don't always capture exactly what was in the author's head. In fact, it is surprising that they work as often as they do.

In using Turing's World, you will have ample opportunity to make mistakes of all three kinds. We only hope that most of the bugs in the design of Turing's

World itself have been eliminated by the time you use it, so that you can make your own mistakes. But we don't make any promises.

1.1.3 Theoretical limitations

Bugs aren't the only headache facing the would-be computer programmer. There are also theoretical limitations on what computers can do. These limitations arise in each of our three steps. In the first place, our ability to model the real world with mathematical objects is limited by our understanding of the world and by the state of mathematics. Novel situations may arise that our model did not anticipate. Or mathematics may not yet contain the concepts needed to model our problem.

Second, in the mathematical domain, it turns out that there are problems which simply don't have any algorithmic solutions of the sort needed by computers. It's not that we don't know how to find such solutions. On the contrary, there are many cases where we know there simply is no algorithmic solution, any more than there is a square circle.

Finally, even if we've managed to model a problem and find an algorithmic solution, we may not be able to implement it on an existing computer, perhaps due to the complexity of the solution. For example, it may require more space than is available on our (or any) computer. Or it might take more time to run than we have: it might take the entire lifetime of the universe, or even longer.

Of these three sorts of limitations, mathematics has the least to say about the first, for obvious reasons. While we can be careful and rigorous about modeling, it is hard to treat the problems inherent in modeling mathematically. Once you have an analysis of some real phenomenon precise enough for mathematics to handle, you've already modeled the real thing in mathematics. Thus computability theory, a branch of mathematics, has the most to say about the last two sorts of limitations. This book does not go very far into computability theory proper. Rather, it introduces you to two of the main notions in the theory, that of a Turing machine and a finite state machine. Our aim is to take readers who know nothing about the theory and, after a few hours playing and working with Turing's World, leave them with a very clear grasp of just what these machines are and how they work.

1.2 Gödel's Theorem

Questions about the theoretical limits of computation were first tackled in the 1930s, by the logicians Kurt Gödel, Alonzo Church, Stephen Kleene, Emil Post, and Alan Turing. While it is Turing's approach that most concerns us, let us place this in some historical perspective. For Turing's work can only be properly understood and fully appreciated from such a perspective.[1]

1.2.1 Hilbert's Program

Gödel, Church, and Kleene weren't actually thinking about computers at all. Even Turing was thinking only of *people* as computers.[2] This should not be too surprising, since there weren't any computers then, at least not what we would call computers today.

What concerned these logicians was trying to understand just what a person could do in a routine or "algorithmic" way, following a step-by-step procedure that could be completely spelled out in advance. They were interested in this problem because a solution to

---

[1]The reader interested in pursuing this further should read the part of Turing's biography [Hodges] about Turing's work on these problems, and his time at Princeton.

[2]By the way, this explains why Turing has his read/write mechanism move, the way human writers work, rather than having the tape move, as one would think of it if it were an electronic digital computer with a paper tape.

it was needed to answer an important question about the nature of mathematics. The question was this. *Can all of mathematics be made algorithmic, or will there always be new problems that outstrip any given algorithm, and so require creative acts of mind to solve?*

In 1920 the great mathematician and logician David Hilbert, flushed with the remarkable achievements of 19th century mathematics, but worried about attacks on its consistency in the early 20th century, sketched a program aimed at providing a new foundation for mathematics. His hope was that the program would save the achievements while settling worries about the consistency of mathematics. The effort became known as "Hilbert's Program."

Hilbert's idea was an important one, even though it did not yield what Hilbert hoped. The idea was this. Mathematics is about numbers, functions, sets, and various other sorts of things, many of them infinite and nonconstructive in various ways. But mathematics is remarkable in that the main activity of the mathematician is that of giving proofs. And proofs, even when they are about infinite objects, are by their very nature finite objects: finite strings of symbols. Hilbert's idea was that we can bring mathematical methods to bear on these strings of symbols themselves, and perhaps in this way show that mathematics is consistent.

Hilbert's suggestion gave rise to a great deal of important research in logic and mathematics. But in 1931 a young Austrian logician, Kurt Gödel, proved two remarkable theorems, now called Gödel's Incompleteness Theorems, which showed that Hilbert's Program could never succeed.

Let's have a shot at describing these theorems, even though it is not crucial for what we are about. From high school math classes, we are all familiar with the idea of Euclidean geometry as the business of proving theorems from some fixed set of axioms by means of certain simple rules of inference. Hilbert's hope was that mathematics would be reducible to finding proofs from a fixed system of axioms, axioms that everyone could agree were true.

One of the main attempts at providing a rigorous axiomization for mathematics had been the system **PM** of *Principia Mathematica*, due to Russell and Whitehead. Gödel took this system as an example and showed how to find true sentences which were not provable in **PM**. Indeed, his proof was so general that it convinced people of a much more powerful claim: *For any set of true axioms and rules of proof that can be spelled out in a determinate way, there are going to be true statements about the natural numbers, 0, 1, 2, 3, ..., that cannot be proven from the given axioms by the given rules.*

The main idea of Gödel's proof is remarkably easy to explain, though the details needed to convert this idea into a rigorous proof are far from easy, even today. Here is the idea. Using the observation that sentences and proofs are finite objects, Gödel showed how to code them by using natural numbers. In this way, he demonstrated how claims about sentences and proofs could be translated into claims about natural numbers. This is not surprising, and was really implicit in Hilbert's program. Here comes the new idea. Gödel was able to construct a sentence about natural numbers which said (under this translation): "I am not provable in **PM**." Let's call this sentence **G**.

Now ask yourself whether **G** is true. Consider first the possibility that **G** is not true. In this case, given what it claims, it must be provable. But intuitively, everything that is provable in **PM** is true. So it must be that **G** is true after all. But then think about what **G** says: that **G** is *not* provable. Thus, it must be the case that **G** is a true sentence that is not provable in **PM**.

Gödel's Second Incompleteness Theorem was even

more devastating to Hilbert's dream. By analyzing the proof of the first result in careful detail, Gödel was able to show that the consistency of **PM** was itself something that could be represented in **PM**, that it was true, but that it could not be proven within **PM**. And again, the proof was so general that logicians could see that it would apply to any similar attempt to axiomatize mathematics. Any consistent system of axioms which could be specified in a determinate way, and which is strong enough to prove certain basic facts about natural numbers, is going to be infected with the same limitation. Its consistency is going to be something that is unprovable in the system. A corollary of the theorem is that there can be no algorithm which will tell you which sentences about the natural numbers are true and which are not. These results completely destroyed Hilbert's dream.

1.2.2 Just what is an algorithm?

While Gödel's results were perfectly correct, they raised a serious problem. The proof was compelling evidence that any attempt to axiomatize mathematics would suffer the same fate as **PM**, but how to state this was completely unclear. To move from the specific case of **PM** and make the more general claim precise, we have to be clear about just what we mean by a "determinate" axiom system.

Roughly, the intuitive idea is that there should be an algorithm for determining, of any given statement, whether it is an axiom of the system in question, and of any given inference, whether it is a legitimate application of the rules of the system. Thus the notion of algorithm actually comes up in a couple of ways in connection with Gödel's theorem: in specifying what would constitute a determinate axiom system in the first place, and then in the corollary, that there is no algorithm for determining what can and can't be proven from the axioms.

We all have a rough idea of what counts as an algorithm and what doesn't. Things like the long division algorithm and the square-root algorithm (which no one remembers any more) are examples. So is the method we all learn for alphabetizing a bunch of words. But getting an answer by rolling dice or asking a mystic are not.

But a rough idea isn't good enough for mathematics. To nail down Gödel's negative answer to Hilbert, in the sense of actually proving what we have claimed Gödel showed, logicians needed a convincing analysis of the notion of an algorithm. It was on this problem that Church, Kleene, and Turing worked. And it was Turing's analysis, using Turing machines, that finally carried the day. Gödel himself, for example, found Turing's analysis the compelling one.

There is an important point, here. A precise analysis of the notion of an algorithm was needed, not to give algorithms to solve this or that mathematical problem. Mathematicians had been developing algorithms, and recognizing them as such, for centuries, without benefit of any precise analysis. An analysis of the notion was needed to understand the *limits* of algorithms, that is, to understand what cannot be done algorithmically at all.

Here is an analogy. Suppose we ask you to construct a triangle with one angle of size 30°, one of size 90°, and one of size 60°. Well, no problem, if you remember your plane geometry constructions. But if we ask you to construct a triangle with angles 30°, 90°, and 100°, what do you do? Try as you might, you fail. But the fault is ours, not yours. We have asked you to do something impossible, since the sum of the angles of a triangle is always 180°. But knowing that fact requires a precise definition of what a triangle is, and a fair amount of work to prove. It took mathematics a long time to get that far.

So too with the notion of an algorithm. If you want to prove that some task cannot be done in a routine, algorithmic way, the way a digital computer in fact works, then you must come up with a precise analysis of the crucial notion of *algorithmic*. Until you've done that, there's no hope of demonstrating its limits.

It is ironic that the work of Gödel, Turing, and others to show the limits of algorithms also showed as a by-product the enormous power of algorithms in connection with Hilbert's original idea, and so led to the development of modern computers.

1.3 Turing machines

Before the work of Turing, there was a great deal of uncertainty about whether the notion of an algorithm could ever be made rigorous. Alonzo Church, building on some work of his student Stephen Kleene, had proposed a characterization of the computable functions, in terms of what has become known as the λ-calculus.[3] However, this characterization was rather opaque and far from convincing at the time it was first proposed.

Turing approached his analysis of algorithms in a novel way. His first step was to think of algorithms in terms of the manipulation of symbols. This idea was borrowed directly from Hilbert's Program: as we've mentioned, Hilbert shifted attention away from mathematical objects to the symbols used to describe those objects. Clearly, if we have some algorithmic procedure, we should at least be able to describe how it works using a finite string of symbols, perhaps by giving a finite list

---

[3]The Greek letter "λ" is pronounced "lambda." The λ-calculus rests behind the modern programming language Lisp and its relatives like Scheme.

of instructions for carrying out the procedure (written, say, in English). What Turing borrowed from Hilbert was the idea that we can also think of the algorithm itself as operating *on* symbols.

The second step in Turing's analysis was to think of algorithms as procedures involving the entirely *deterministic* manipulation of symbols. For this he imagined the algorithm being carried out by a deterministic device—a machine of some sort—that operates on symbols written linearly on an arbitrarily long tape. He suggested that we think of these as simple physical devices that can read or write only one symbol at a time on the one-dimensional tape. For simplicity, we can imagine that the tape is marked off into squares, each with room for just one symbol, and that at any given time the machine is surveying exactly one of the squares.

Now it would be easy to go astray at this point and start worrying about what the inside of such a device might be like. But Turing had the genius to realize that he didn't have to assume anything at all about the physical nature of the devices. To guarantee that the procedures carried out by such devices are really algorithmic, we need only assume that they operate according to some fixed list of instructions. To capture this intuition, Turing required that each such device have only a finite number of possible "states," and that at any given time the device must be in exactly one of those states, the state corresponding to the instruction it is about to carry out.

The next question is what sorts of actions the device can perform, what kinds of things can the instructions instruct it to do? Turing broke down the various things the device might do into a very small number of basic operations. The device reads a single square of the tape and then does one of three things, depending on what it sees and the state it is in: it can write some symbol

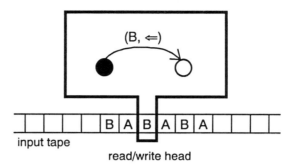

input tape

read/write head

Figure 1 A Turing Machine

on the square (replacing whatever it saw), or move one square left on the tape, or move one square right. After carrying out one of these actions, the machine goes into some state, possibly different, possibly not. That's all.

Let's summarize. The devices Turing imagined, Turing machines, as they've come to be called, each has a finite number of possible states. At any given point, such a machine is surveying a single square on a long, thin tape. What the machine does at that point—whether it writes a symbol, moves left or moves right, and what state it will then go into—is completely determined by the state it is presently in and what it presently sees in the square.

It is customary to picture a Turing machine as in Figure 1. This machine is pictured over a tape on which some A's and B's are written; its read/write head is poised above a B. Inside the machine we "see" two "states," the dark one being the state the machine is in, the light one the state it is going into. The label (B, ⇐) on the arc between them indicates that when the machine is in the state on the left and sees a B, it should move one square to the left.

Despite the conceptual simplicity of these machines, Turing conjectured that a symbolic procedure is computable or algorithmic, in the intuitive sense, just in case we can design such a machine to carry out the procedure. This claim is known as *Turing's Thesis*. It is a striking thesis, and at this point may seem rather implausible: could it really be the case that every algorithm can be performed on some, suitably designed Turing machine? But in spite of its initial implausibility, it is now almost universally accepted. To see why, though, you first have to convince yourself just how powerful these simple machines really are. And that is just what Turing's World is for, to let you come to understand how these simple machines can do amazing things when constructed in the right way. Amazing things, but not just anything: there are limits to the power of computers, and that is what computability theory is really about.

1.4 Other machines

Since the publication of Turing's original paper, his work has been generalized by many people in many different ways. One important line of research explored alternative architectures for computing machines, architectures where the machine is conceived of in quite different ways. But it turns out that none of these competing architectures is more powerful than Turing's: every algorithm that can be computed on any of these alternative machines can also be computed by a Turing machine. This is part of the evidence for Turing's thesis. Turing also showed that the Turing computable functions are the same as those definable in Church's λ-calculus, thus giving compelling evidence for Church's thesis that the λ-definable functions are exactly the computable functions.

We will not investigate these alternative architectures or the λ-calculus. There is, however, another line of generalization of Turing's work that we will pursue.

It is possible to take the basic idea of a Turing machine and modify it, either by limiting the kinds of things it can do, or by adding new capabilities. Turing's World allows you to explore several of these.

We mentioned earlier that Turing machines operate in an entirely deterministic manner. What the machine does at any given time is completely determined by the symbol it is scanning on the tape and the state the machine is in. An important generalization is to consider "nondeterministic" Turing machines, machines that may perform various different actions in a given state and scanning a given symbol. It turns out, rather surprisingly, that any algorithm that can be computed by such a machine can also be computed by some other, deterministic Turing machine, though the latter may take a lot more time.

Nondeterministic Turing machines have a capability not possessed by regular Turing machines. But what if we take away some of the capabilities? In particular, what if we do not let the machine change the tape at all? One might think that such machines would be totally pointless, but this has turned out to be far from the case. These machines, called "finite state machines," are of considerable interest in computer science due to their computational efficiency.

Turing's World, besides giving you the power to explore Turing machines, also lets you experiment with various types of finite state machine, as well as with nondeterministic versions of both Turing machines and finite state machines.

1.5   Further reading

In this section we present an annotated guide to a few of the many works on Turing machines, so that the user whose appetite has been whetted by Turing's World can pursue their study. The list here is far from complete.

References

Barwise, Jon, and John Etchemendy. *The Language of First-order Logic* (including the program *Tarski's World*), CSLI Lecture Notes 23, 1992, 259 pages.

This is a beginning textbook on first-order logic. It comes with a computer program to help teach the basic semantic notions of logic. Later in this book, we will present some exercises that relate computability theory and logic, and readers unfamiliar with basic propositional logic may want to consult this or some other introductory text before tackling these exercises.

Boolos, George, and Richard Jeffrey. *Computability and Logic*, 3rd ed, Cambridge: Cambridge University Press, 1989, 304 pages.

An undergraduate text on computability and logic that begins with a discussion of Turing machines. The format of the state diagrams used in Turing's World is designed to be compatible with this text. The book works through a thorough discussion of Gödel's Incompleteness Theorems and has a chapter on the Busy Beaver function, which we explore in Exercises 42–46.

Barwise, Jon. *Handbook of Mathematical Logic.* Amsterdam: North Holland, 1977, 1165 pages.

A large reference work in mathematical logic, about one-quarter devoted to computability theory and its applications. Enderton's chapter (Elements of recursion theory) presents a quick and excellent introduction to the basic results and starts off with the notion of Turing machine. It uses the 5-tuple approach that we discuss in Exercise 33, page 98. Applications follow in the chapters by Davis (Unsolvable problems) and Rabin (Decidable theories). Gödel's Theorems are discussed in the chapter by Smorynski.

Davis, Martin. *Computability and Unsolvability.* New York: McGraw-Hill, 1958, 210 pages. (Also available from Dover.)

A short book containing a detailed account of computability theory entirely based on Turing machines. It does not presuppose any particular mathematical background. On the other hand, it does require more in the way of a taste for mathematical style and symbolism than the other references listed here. Of the references given here, it contains the most thorough discussion of Turing machines. Like [Boolos & Jeffrey], it uses the 4-tuple approach.

Haugeland, John. *Artificial Intelligence: The Very Idea.* Cambridge, Mass: Bradford Books, The MIT Press, 1985, 224 pages.

A popular but very sound introduction to artificial intelligence, its foundations, promise, and problems. It is quite readable and has a nice informal chapter on Turing machines. This includes a description of a very small (and inscrutable) universal Turing machine that is due to Marvin Minsky, one of the founders of AI.

Herken, Rolf. *The Universal Turing Machine: a half-century survey.* New York: Oxford University Press, 1988, 661 pages.

A collection of papers assessing and extending Alan Turing's work, written in commemoration of the 50th anniversary of the publication of [Turing]. Of particular interest to the nonspecialist is Robin Gandy's contribution, which discusses the historical context in which Turing's work arose.

Hodges, Andrew. *Alan Turing, the Enigma.* New York: Simon and Schuster, 1983, 576 pages.

An interesting and well-written biography of Turing. It contains a very readable discussion of Turing machines and gives a nice sense of the historical setting in which they arose, and of the way Turing himself thought of them and their relationship to minds.

Hopcroft, John. "Turing Machines," *Scientific American,* May 1984: 86–98.

A popular, but comprehensive presentation of Turing machines and their relationship to a number of the important problems in computability.

Hopcroft, John, and Jeffrey Ullman. *Introduction to Automata Theory, Languages and Computation.* Palo Alto: Addison-Wesley, 1979, 418 pages.

A more advanced textbook than [Boolos & Jeffrey]. While it does not go into Gödel's Theorems, it does focus more on the symbolic aspect of computation. It begins with a discussion of finite state machines and works up to the theory of Turing machines in Chapter 7. The text also discusses the P=NP problem, which we touch on in Exercise 69, page 115.

Kleene, Stephen C. *Introduction to Metamathematics.* Princeton, N.J.: Van Nostrand, 1952, 550 pages.

This is the classical reference on recursion theory and Turing machines (Part III of IV), plus a great deal more in logic. For many years it was the only textbook on the subject, prompting many to remark that Kleeneness was next to Gödelness. It contains a detailed proof that the recursive functions are exactly those that are Turing computable, a topic we cover in Exercises 34–41, starting on page 99.

Turing, Alan M. "On computable numbers, with an application to the Entscheidungsproblem," *Proc. London Math. Soc., Ser.2*, **42**: 230–265.

Turing's original paper, in which the notion of a Turing machine is first introduced and the most fundamental facts about them established. It is a classic paper in mathematical logic and inspiring reading even today.

Whitmore, Hugh. *Breaking the Code.* New York: Samuel French, 1988.

This highly successful play about the life and work of Turing (based on [Hodges]) starred Derek Jacoby when it played on Broadway. It has excellent, short explanations of Turing's basic ideas on Gödel's work and of how he developed the notion of a Turing machine.

2

Running Turing Machines

In this chapter we introduce you to the basics of Turing's World by guiding you through the use of some prepackaged Turing machines. Besides illustrating some features of Turing's World, this will give you a chance to see a few Turing machines in action.

Recall that Turing machines operate on input tapes that come marked off into squares. Each square contains a single symbol. Any particular Turing machine is a finite device whose behavior is completely determined by the symbols on the tape. One step in a computation has the machine scan a square and then carry out a single action depending on what symbol it has seen. The range of possible actions is severely limited: it can print some symbol (including, perhaps, a blank), it can move left one square, or it can move right one square.

There are many ways of representing Turing machines: state diagrams (sometimes called "flow graphs"), sets of 4-tuples, and machine tables are all common. Turing's World uses state diagrams as the primary mode of representation, but also allows the use of 4-tuples (which we discuss in Chapter 6). State diagrams are basically just flow charts and so are the easiest way to specify a Turing machine. Your Turing's World program uses the state diagram you draw to simulate a real Turing machine.

Because the theory of Turing machines is usually taught within the context of number-theoretic compu-

tation (computation on the natural numbers $0, 1, 2, \ldots$),
a fundamental aspect of Turing's approach is often obscured: the use of symbolic computation. Any old symbols are as good as any others in Turing's World. To
emphasize this, most of the examples we discuss in this
book involve symbolic computation. In particular, all
of the Turing machines we consider in this chapter operate on strings made up of the four characters: A, B,
(,), plus the blank.

Consider this problem from "algebra." Given some
expression made up of these four symbols, check to see
if it is well formed in the sense that all its left and right
parentheses match. If so, simplify the expression by
deleting the parentheses and grouping all A's before all
B's. In this section we present three machines for doing
pieces of this task, and then later we will show you how
to hook these three together into one large machine to
carry out the whole task.

The first of our prepackaged Turing machines is designed to take a string of A's and B's, and rearrange
them so that all the A's come first. The second machine
is designed to take a string that includes parentheses:
it deletes all the parentheses and contracts the result
on the tape, so that there are no spaces in the middle
of the string. The third machine examines a tape that
contains a string made up from the four characters and
tests to see if the parentheses are properly matched. If
they are it prints GOOD, if not it prints BAD.

A word is in order about the notation in our state
diagrams. Suppose we want to have a machine that acts
as follows: if it is in state 17 and sees an A then it moves
right and goes into state 2. This action is represented in
a state diagram by having an arc pointing *from* a node
labeled 17 *to* the node labeled 2; this arc itself will be
decorated by the label $(A, \Rightarrow)$, the A for "if you see an A"
and the $\Rightarrow$ for "then move right." If we had labeled the
arc (A,B) then it would have meant that if the machine

sees an A it should print a B; similarly, (A,–) means that
if the machine sees an A it will "write" a blank, that is,
erase the A.

A machine to reorder a sequence

Our first Turing machine is designed to perform a very
simple task: take an arbitrary string of A's and B's and
rearrange it so that all the A's precede all the B's. We'll
call this machine the Lone Rearranger.

2.1   Starting up Turing's World

We assume you have a rough idea of how to find your
way around on a Macintosh. If you are new to the Macintosh, glance over the Appendix on page 117, which
explains some of the standard Macintosh terminology,
such as *menu* and *double-click.*

There are a couple of ways to start up Turing's
World, depending on whether you plan to immediately
run a machine or to build a new machine. Here we'll
explain the former way; the latter is explained in Chapter 3.

Turn on your Macintosh and wait until the Finder
appears. The Finder is the initial Macintosh program
from which you launch other programs, move and delete
files, and so forth. You'll recognize it by the trash can
icon in the bottom-right corner of the screen. Insert
your Turing's World disk into the disk drive. In a moment, a new icon will appear toward the right of the
screen, representing the Turing's World disk.

To see the contents of the Turing's World disk,
double-click (that is, click twice in quick succession) on
the Turing's World disk icon. A window will open that
contains two items: the Turing's World program and a
folder called Machines and Tapes. We want to open a
machine that's been saved inside this folder, so double-

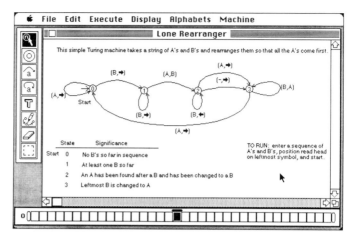

Figure 2 The Screen

click on the folder. Another window will open, showing icons for various Turing machines.

Look for the icon titled **Lone Rearranger**, since that's the machine we want to run. Double-click on this icon. This will launch Turing's World and open the state diagram for the **Lone Rearranger**. In a moment, your screen will look something like Figure 2.

At the top of the screen you see the usual Apple, as well as six menu labels: **File, Edit, Execute, Display, Alphabets** and **Machine**. Pull down each of these menus to see the sorts of things they contain. At the far left of the screen is the **Tool Palette**, with eight buttons. We won't be using these until the next chapter.

The Turing machine's tape is represented at the bottom of the screen, with the scanned square (the location of the read-write head) highlighted in black. The large window to the right of the tool palette is where you design Turing machines by drawing state diagrams of them. Since you've opened the Lone Rearranger, its state diagram now fills this window and the words "Lone Rearranger" appear in its title bar.

Notice that we've annotated this state diagram with some information to help you understand how the Lone Rearranger works. But the English expressions themselves have no significance for the machine's actions. They are there simply for our benefit in understanding the machine, or for remembering what we had in mind when we designed it. (At this point, though, don't worry about whether you understand how the Lone Rearranger is supposed to work.) Whenever you design a Turing machine you should add similar annotations, both for the benefit of others, and for your own benefit, later on.

2.2 Opening a tape from a file

To see this machine in action, we need to get some A's and B's on the tape. In a moment, we'll explain how to write directly on the tape, but for now we will use a prepackaged tape. Pull down the **File** menu and choose **Open Tape....** A list of the tapes in the current folder (Machines and Tapes) will appear. We want to open Tape 1. To do this either select the name Tape 1 on the list and then click the **Open** button, or simply double-click on the name Tape 1. This will load a sequence of A's and B's on the tape, with the read-write head (the highlighted square) on the leftmost symbol.

2.3 Running a Turing machine

Now pull down the **Execute** menu. The first five items on this menu allow us to run the machine at various speeds: graphic, fast, lightning, step, and reverse. Very briefly, these operate as follows:

Graphic: This carries out a computation at a slow enough pace that you can watch its action graphically displayed on the screen.

Fast: This is just a much faster version of the above. It goes so fast that you will not be able to follow the computation in detail.

Lightning: In this mode, the computer carries out the computation in secret, as fast as possible, and then displays the result on the screen.

Step: This allows you to step through a computation one move at a time, taking as much time between steps as you want.

Reverse: This allows you to back up a computation, one step at a time.

Let's try out some of these on the Lone Rearranger. Start by choosing **Graphic Run**. This first time through, watch the read-write head as the machine rearranges the tape. You see that the machine works its way right along the tape until it encounters an A after a B. It changes this A to a B, then backs up to the leftmost B, which it switches to an A. It then starts right again, going through this whole routine, until it makes a successful pass, one where no A is found after a B. The beep means that the machine is finished, that it has "halted."

Let's run the same tape again, only this time watch what's going on in the state diagram. Open Tape 1 again (from the **File** menu) and start the machine by choosing **Graphic Run** again. Each arc on the diagram represents a possible action of the machine. For example, the arc from node 0 to node 1, labeled (B,⇒), indicates that when the machine is in state 0 and sees a B on the scanned square, it should move right and go into state 1. When the machine carries out this action, Turing's World lets you know by briefly highlighting the nodes and blinking the arc. This highlighting bounces around as the machine moves from state to state, giving you a graphic representation of the machine's actions.

Let's try rerunning this machine at two of our other speeds. Reopen Tape 1, and choose **Fast Run** from the **Execute** menu. Here the machine will execute as fast as your Macintosh can redraw the screen after each step. Finally, open Tape 1 again and try running the machine by choosing **Lightning Run**. This time the machine executes without displaying anything except the final state of the computation.

These different speeds have different uses. Of the three, graphic is usually the best for understanding exactly how a machine is working. But even then you really need to keep one eye on the state diagram and one eye on the tape. And that's not easy, except maybe for frogs. So let's look at a simpler machine, and at how to slow the computation down even further.

2.4   Opening a machine from a file

The next machine we'll look at just consists of the first two nodes of the Lone Rearranger: it proceeds down the tape until it finds the first A that appears after a B. We've called it Simple Simon.

To open a new machine from a file, pull down the **File** menu and choose **Open State Diagram**.... When you do this, you will be shown an alphabetized list of the state diagram files saved in the current folder (Machines and Tapes). At first, you won't be able to see Simple Simon on this list, since it is at the bottom of the list. To find the name, scroll the list by clicking on the downward pointing arrow in the scroll bar to the right of the list. When Simple Simon appears, open it the same way you opened Tape 1: either double-click on the name Simple Simon, or select it and click **Open**. (If you have accidentally made some changes to the Lone Rearranger, you will be asked if you want to save the changes. Just click **No**.)

This time, let's open Tape 2: recall, you open tapes

from the **File** menu by choosing **Open Tape**.... Before running Simple Simon, though, read the next section.

2.5 Stepping through a computation

To see exactly how a Turing machine runs, it's helpful to move step by step through a few computations. If you choose **Step** from the **Execute** menu, the machine will carry out exactly one step in the computation and then pause. By "one step in the computation" we mean the machine will make one transition along an arc. It won't necessarily change from one state to another. By choosing this command repeatedly, you can step through the entire computation.

There is an easier way to do this. Instead of choosing the command **Step** from the **Execute** menu, you can do it from the keyboard, by holding down the command key (it has a cloverleaf on it) and typing "E." This is called a command key equivalent for the menu item, and is written "Command-E." Most of the menu commands have a command key equivalent; these are indicated on the right side of the menu. For example, you could have carried out a graphic run by typing Command-G.

Try stepping through the computation by repeatedly typing Command-E. See if you can predict each time what the machine will do next. Keep stepping through until you hear the beep that indicates the machine has halted. Then try it again: reload Tape 2 and step through the computation once more. Do this until you're confident you understand how Simple Simon works.

Exercise 1 Open Tape 3 and step through the computation. What state does it halt in? Why? Were we right when we said Simple Simon moves down the tape until it finds the first A after a B? What would be a better description of its behavior?

2.6   Resetting the machine

In the middle of a computation, whether you are stepping through or on a continuous run, most of the menu
items are "greyed out." If you want to do something
else without carrying out the computation to the end,
you must choose **Reset Machine** from the **Execute**
menu (or type Command-M). This will stop the computation and put the Turing machine back into its starting
state.

At this point, you should open the Lone Rearranger
again and step through the computation of Tapes 1 and
2. See if you can figure out exactly how the machine
works. In particular, notice how the machine manages
to find the leftmost B to change it to an A. At some
point, reset the machine in the middle of the computation, just to try out this command. Also note that
you can start a continuous run after stepping through
part of a computation, by choosing **Graphic Run,**
Fast Run, or **Lightning Run** from the **Execute**
menu.

2.7   Writing directly on the tape

You are probably getting tired of the prepackaged tapes,
and would like to see if the Lone Rearranger works correctly on tapes other than the ones we've provided. To
write directly on the tape, you first need to position the
read head where you want your symbols to start appearing. You can position the read head by using the
left arrow ($\leftarrow$) and right arrow ($\rightarrow$) keys on the keyboard. These will move the read head in the expected
directions.

When the read head is where you want it, you can
type directly on the tape from the keyboard. At this
point, the only keys that will affect the tape are "A,"
"B," the space bar, and the delete (backspace) key. If

you type any other key, your Mac will beep at you, indicating you have tried to type something not in the alphabet used by this machine.

Try typing on the tape. Don't worry about changing it, since the changes won't be saved unless you ask for them to be. Create a tape for use with the Lone Rearranger: some random sequence of A's and B's with no spaces separating them. Then move the read head to the leftmost symbol using the arrow keys. You are now ready to run the machine on your own tape.

Exercise 2 Run the Lone Rearranger on several different tapes that you create. Check it on tapes that are all A's or all B's. On one of the tapes, try starting it with the read head in the middle of the sequence of symbols. Under what conditions does the machine produce the wrong results?

You can also modify the tape using the mouse, rather than the keyboard. You can move the read head by clicking down on it and then dragging it to the desired square, letting up the mouse button when you get there. Try this out.

To change what is written on a square of the tape using the mouse, first choose **Show Alphabet** from the **Alphabets** menu. This gives you a small window showing all the symbols that can be read by the current machine. One of them will be highlighted. If you click on a square of the tape, this symbol will appear. If you click again, this symbol will be replaced by a blank. You can change which symbol is highlighted by clicking on the symbol in the alphabet window.

At this point we encourage you to play around a bit, both so you get used to making tapes and running the machine, but also to explore the behavior of our simple, sample machine. For instance, try starting the machine with the read head on a blank square. What does it do?

Is that what it "ought" to do? What happens if there is a blank in the middle of your sequence? Why?

These are important lessons to be learned about Turing machines. But while playing around, you might also discover some things about the Turing's World program itself. For example, you might notice the little number at the left of the tape, and that it changes as the Turing machine runs. What does it indicate? Or you might discover that you can drag the alphabet box around by clicking on or near the title **Alphabet**. And you get rid of the alphabet box by clicking the little square in its upper left corner.

2.7.1 Typing logic symbols

If your alphabet contains the logic symbols ¬, ∧, ∨, →, ↔, ∀, and ∃, you can type these symbols directly from the keyboard using the following table of keyboard equivalents. Note that many of them require two keystrokes. For example, to type ¬, you must type Option-u Shift-u, that is, type "u" twice, first holding down the Option key and then holding down the Shift key.

Symbol	Keyboard Equivalent
¬	Option-u Shift-u
∧	Option-e Shift-e
∨	Option-o
→	Option-u Shift-o
↔	Option-n Shift-n
∀	Option-a
∃	Option-s

2.7.2 Clearing the tape

To clear the tape, choose **Clear Tape** from the **Execute** menu. This replaces all the symbols on the tape with blanks.

2.8 Getting frozen: when nothing works

Sometimes you may find that nothing you want to do will work. You may try to write on the tape (or add to a machine you are building) but Turing's World does not respond. Typically this is because you are in the middle of an uncompleted computation. Perhaps you have paused, or were stepping through a computation which you never finished. To get out of this situation, simply reset the machine by choosing **Reset Machine** from the **Execute** menu. The only other situation in which this may occur is if you've forgotten you are in the middle of labeling an arc, a procedure we explain in the next chapter.

2.9 Seeing more of the tape

One important feature that you might stumble upon is that the tape is really much longer than it looks. Click one of the tiny rectangles at either end of the tape and the tape will slide one square in that direction, with more squares appearing at the other end. Your symbols, along with the read head, will be pulled along as you click. Notice, though, that the read head must always stay visible, so if you shift the tape past where the read head is sitting, the read head gets left behind. Try this feature out.

When we reviewed the definition of a Turing machine, we said that the machine's tape was "arbitrarily long." People often describe the tape as being infinitely long. It doesn't really matter how you describe it: what is important is that a Turing machine must never run out of tape in the middle of a computation. Whether you use an infinitely long tape or simply station extraordinarily productive paper mills at each end makes no difference. It turns out that this assumption, that we'll never run out of tape, is in fact a very important idealization, one that accounts for much of the power of Turing machines.

2.10 Quitting

Eventually you will want to turn off Turing's World and do something else. Or maybe something serious has gone wrong while you were fooling around, and you want to start the program over. To do this, choose **Quit** from the **File** menu (or type Command-Q). You may be asked if you want to save changes to the state diagram (if you accidentally made any); just click **No** unless you really do want to save them.

A machine to remove parentheses

Of course if you want to go on to the next machine, you don't have to quit and start over. From the **File** menu, open the state diagram called Paren Remover.

At first sight, the Paren Remover looks similar to, though much simpler than, the Lone Rearranger. But appearances can be deceiving. Notice that the nodes of the diagram are square, not round. Turing's World allows you to take a whole Turing machine, shrink it to a submachine, and use that as a single node in a bigger machine. We've taken advantage of this feature in building our parenthesis remover.

Before running this machine, take a look at the alphabet (**Show Alphabet** from the **Alphabets** menu). You will discover that in building this machine, we've used an extra symbol (the number sign) in addition to the symbols A, B, (,). This is a common trick: using one or more auxiliary symbols often simplifies a Turing machine, for reasons that will soon be apparent. But it turns out that these extra symbols can always be gotten around in principle, though usually at the cost of much more complicated machines.

Put a fairly long sequence of A's, B's, ('s, and)'s

on the tape (at least 25 characters long) and position the read head at the left. (Or, if you'd rather, open the prepackaged Tape 4.) To get an overall idea of how the Paren Remover operates, we'll make the first run at the fast speed. Select **Fast Run** from the **Execute** menu. Do this again, this time watching the machine's diagram, as it moves from submachine 0 to 1 and on to 2. Are the submachines behaving as described on the state diagram?

2.11 Looking at a submachine

To see how the various submachines are designed, we will examine their state diagrams. Make sure the top button on the control panel, the one showing the picture of a node with an arrow pointing at it, is highlighted in black; if it's not, select it. Then double-click on one of the square nodes representing a submachine. A new window will open up, one that shows the state diagram of that submachine. If you run the machine on Fast Run with one of these submachine diagrams in the top window, that will be the one you will see when the machine does its thing.

Now close the submachine and run the machine on a shorter sequence, but do it using **Graphic Run**. At this speed, the various submachines will be displayed as they are used. Similarly, they are displayed when you step through a computation. Since you'll eventually want to build machines with submachines yourself, you might want to step through a couple of computations to get a feel for what is going on.

2.12 Pausing and changing speed in mid-run

You may have noticed that you can change the speed of the execution in the middle of a computation. If the machine is running and you pull down the **Execute** menu, things come to a temporary halt, and then you can select a new speed. (You can also do this with command key equivalents.) If you choose **Pause** (or type Command-P), the machine will pause and won't start up again until you tell it to by choosing one of the run commands. Also, at any point during a continuous run you can begin stepping through the computation by typing Command-E (or choosing **Step**).

2.13 Computational complexity

In Chapter 1 we discussed the fact that the complexity of a given algorithm may for various reasons outstrip what we can actually implement on real computers. An important part of theoretical computer science concerns issues of computational complexity, questions of how complicated a given machine or computation is. Complexity can be measured along various dimensions: number of states, number of symbols, how much tape is used in a computation, and how many steps a computation takes. There are trade-offs in all these dimensions.

How many steps do you think it would take to remove the parentheses from the string (A(B))A()B? There are only six parentheses, and ten symbols in all. Our parenthesis remover is pretty inefficient, though. That is one price of the modularity we've built into it with the three separate submachines. To see how many steps it takes, run the machine on this string, and then choose **Time-space count** from the **Display** menu (or type Command-I). You'll see that the computation took 120 steps, but used only 12 tape cells. Later you may want to design a parenthesis remover that cuts down the number of steps as much as possible.

A machine to check parentheses

Our next prepackaged Turing machine checks a string made up of A's, B's, ('s, and)'s to see if the right and left parentheses are properly matched. Formally, the set of *properly matched strings* is defined as follows:

1. If α is any string containing no parentheses, then it is properly matched.
2. If α is a properly matched string, then so is the string: (α).
3. If α and β are properly matched strings, then so is the string $\alpha\beta$.
4. Nothing is properly matched except in virtue of 1–3.

Thus, for example, ABA and BBA are properly matched, by (1). By applying (2) to each of these, we see that both (ABA) and (BBA) are properly matched. Applying (3) we see that (ABA)(BBA) is properly matched. Applying (2) again, we see that ((ABA)(BBA)) is properly matched. On the other hand, (AAB)(BBA)) is not properly matched.

Is () properly matched, according to our definition? It depends on whether a "string" can be empty, whether it can contain no symbols at all. If so, then () results from putting parentheses around the empty string, which is properly matched by clause (1), and so () is also properly matched, by clause (2). Throughout this book we will consider empty strings to be legitimate strings, unless otherwise indicated.

You should now have no trouble opening the machine called Paren Check and writing a sequence for it to compute. But before you run it, there are a few new things that this machine illustrates.

2.14   Seeing more of the state diagram

The parenthesis checker is a fairly large machine. While the main diagram has only ten nodes, eight of those nodes represent submachines. And even with the ten main nodes, there is not enough room on the screen provided by a small Macintosh, so the machine extends off the page in two directions. If you have a larger screen, you can resize the window by dragging the resize box in the lower right corner of the window down and to the right. Otherwise, you can see the portion of the machine on the right by clicking a few times on the scrolling arrow at the lower right corner of the state diagram window. (The arrow you want is the one that points right.) To move back, click the left scrolling arrow at the other end of the scroll bar. You can see the rest of the text describing the machine by clicking on the down scrolling arrow at the bottom of the vertical scroll bar.

2.15 Saving design time with submachines

If you examine the various submachines used in the parenthesis checker, you will discover that many of them are variations on a theme. For example, several of them simply move right looking for some symbol or other. It would have been extremely tedious to have to draw each of these separately. As you'll see, Turing's World lets you copy a given submachine to use in another place, either exactly as is, or with modifications. We've used this feature repeatedly in designing this machine. We will show how to do this in the next chapter.

Exercise 3 Run the Paren Checker on several different tapes that you create. You should start the runs with the read-write head on the leftmost symbol of your input string. Make sure to check it on a tape containing the string (). Does it give the right answer? How about if you run it on the empty string, that is, on a completely blank tape?

Exercise 4 Open the state diagram called Monadic Multiplier, and run this machine on a tape containing 3 *'s, followed by a space, followed by 4 *'s (starting the read head on the leftmost *). When it is done, count how many *'s are left on the tape. Run the machine on other, similar tapes. See if you can understand what it does and how it does it. This machine is based on a design presented in [Boolos & Jeffrey].

3

Building Turing Machines

It is high time we explained how to build a Turing machine using the environment provided by Turing's World. In this chapter we will lead you step by step through the construction of a simple machine.

3.1 Getting ready to build a machine

To build a machine, you need a blank state diagram window to work in. If you start Turing's World by double-clicking on the Turing's World icon (rather than the icon for a saved machine) at the Finder, you are automatically given a blank state diagram window. But if you've been using Turing's World already and have a state diagram in the main window, choose **New State Diagram** from the **File** menu (or type Command-N). This will give you an empty window titled "Untitled State Diagram," all set for you to go to work.

3.2 Using the default alphabet

As we've seen from the previous chapter, Turing's World allows you to design machines with fairly rich alphabets. But one special alphabet is used for most theoretical work. This alphabet has only the blank and *, and is called the *default alphabet*. You have already seen one machine that uses the default alphabet, the Monadic Multiplier from Exercise 4. The first machine we build will use this alphabet as well.

Pull down the **Alphabets** menu, and make sure that **Default Alphabet** is checked. If not, select it.

3.3   Using the tool palette

Along the left side of the screen you will see the *tool palette*, used for drawing and annotating state diagrams of Turing machines. It consists of eight buttons in a vertical column. Selecting one of these items allows you to do something specific in the state diagram window, the portion of the screen currently labeled "Untitled State Diagram."

The first button is used to put down nodes representing the possible states of your machine: it is the one currently highlighted in black. The second button allows you to change the nodes to a special kind (accepting nodes) that are used in finite state machines, which we will discuss later. The third and fourth allow you to draw arrows or "arcs" between nodes. The only difference between these two is in the shape of the arc. Some people prefer one, some prefer the other, some like to mix the two. The node and arc buttons are all that are needed for the actual design of a Turing machine.

The fifth button, the one labeled **T**, is used when you want to annotate your Turing machine. The "**T**" is for "Text." The sixth button, the one representing a drawing pencil, is used to redraw the screen. This is useful since editing sometimes leaves the screen in a state that is hard to read. The next to last button, representing an eraser, is used to erase the state diagram completely, in case you want to start over again. Whenever you click this, you will be given a chance to back out, in case you didn't really want to erase everything. The final button is called the selection tool, and is used to move, copy, or delete a portion of a state diagram.

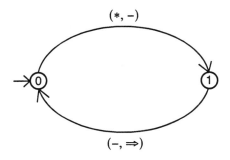

Figure 3 The Eraser

A machine to erase *'s

We're going to build a simple machine that we call the Eraser. It just erases the first bunch of *'s it sees to the right and then halts when it gets to a blank. There are a lot of ways to build a machine that does this. The one we'll construct has the state diagram shown in Figure 3.

If things go wrong at any stage, just clear the screen (by clicking the eraser button on the tool palette) and start over. Later we'll show you how to edit your diagrams so you won't have to start over when any little thing goes wrong, or when you change your mind about something.

3.3.1 Dropping state nodes

The first thing you need to do is to put down a couple of nodes to represent the two states the machine can be in. To do this, click the **Node** button on the tool palette, to make sure it is active. The active item is always highlighted in black. Now move the mouse to some spot toward the center of the state diagram window and click. This will drop your first node, and label it 0. Now move the mouse a couple of inches to the right and click again.

There, the two nodes, representing the two states of the Eraser, are in place. If at any time you want to add some more nodes, just select the **Node** button

again, and drop some more. The labeling is taken care
of automatically. (Sometimes the numbering is not the
one you might choose, but the numbers don't have anything to do with the way the machine operates. They're
only for us, to make it easy to talk about the machine's
operations.)

Notice that the first node you created, node 0, has
an arrow next to it. This is to indicate that state 0
is the machine's initial or starting state. Later, we will
explain how to change the start state to one of the other
states.

3.3.2 Drawing arcs

Now it's time to draw and label the arcs. Each arc
points from a *source* node to a *destination* node. To
draw one, select one of the **Arc** buttons on the tool
palette. Move the mouse to node 0. When it is on
top of the node, a little src will appear above the arrow.
This stands for "source." Click once. An arc with its tail
attached to the source node will appear. Its other end
will be attached to a cursor in the form of crosshairs. As
you move these over nodes, a dst will appear, standing
for "destination." Move this over node 1 and click. The
node is now anchored at both ends.

3.3.3 Labeling arcs

Each arc label has two parts, an *input character* and an
action. Turing's World will help you label your arc by
asking you two questions: what is the input character
and what is the action? These questions will appear
in a window at the lower right of the screen. In the
present case, you want to label your arc (∗,–), meaning
that if the machine sees a ∗ it should erase it, i.e., print
a blank. The window contains the question: "Input
character = ?" This label (∗,–) has the input character
∗ so you should answer by clicking the ∗. Next the
question "Action = ?" appears. The possible answers

are displayed below the question. You can either write *
or blank, or move left or right. In our case, the correct
answer is to print a blank, so you should click the bl
button.

If you accidentally choose the wrong button for the
input character, you can start over by selecting the Cancel button next to the input options. You will have
to draw the arc over again after doing this. When selecting an input character, you will notice a button labeled otherwise above the Cancel button. Stay away
from this button for now; we'll explain what it does in
Section 6.2.

3.3.4   Positioning arcs

Now your arc is labeled. All you have to do is position
it. Of course you may be happy with it where it is, but
for practice, you should move it somewhere, say a bit
higher. To do this, move the cursor over the blinking dot
(or over the comma in the arc label), until it turns into
an X. Then drag the label where you want it. (Whenever
you know that there are going to be two arcs with the
same source and destination, you should move all but
the last out of the way as you create them, so that the
arcs don't end up obscuring one another.)

Draw, label and position the second arc. This time
node 1 is the source and node 0 is the destination. The
label should read $(-, \Rightarrow)$. Your machine is now complete.

Warning: A common mistake in building machines
is to forget to select an **Arc** button when you want to
draw an arc, instead leaving the node tool selected. If
you click on top of an existing node with the node tool,
the node will start blinking. This is for use in editing.
For now you can simply move away from the node and
click, to restore stability.

Tip: There is an easy way to move between the top
four tools on the tool palette, using a combination of the

shift key and the option key. If you have either node tool selected, the option key will move you to an arc tool, and vice versa. The shift key moves you from one node tool to another, or from one arc tool to another. Thus by holding down a combination of the shift and option keys, you can get from any of the first four tools to any of the others. Try this out, for future use. You will notice that the tool reverts to its original position when you release the keys, so you must hold them down while you use the tool selected.

3.3.5 Annotating a state diagram

You should immediately annotate your state diagram. You can write anything you want, since it won't affect the operation of the machine. This little machine is so simple that it may not seem worth the trouble to annotate it, but in the long run it is a good idea to annotate every machine with enough information to allow you or anyone else to use it without having to figure out what it is designed to do.

Suppose we want to add the comment: This machine erases an initial sequence of *'s and then halts. Select the **T** button on the tool palette. Now click on the screen where you want your text to start. Just type in your text. When you finish, either click somewhere else in the state diagram window or select another tool from the palette. You can edit the text using most of the standard Macintosh editing techniques. We will describe editing techniques later.

Exercise 1 Add text explaining how to run your machine.

3.3.6 Redrawing the screen

Clicking on the pencil button simply redraws the screen. You can do this any time things don't look quite right.

3.3.7   Clearing the screen

Clicking on the eraser button clears away your work, allowing you to start over. Don't worry, though. It gives you a second chance in case you weren't really serious about throwing it all away. Try it and then click **Cancel.**

3.3.8   The selection rectangle.

The last tool on the palette is the selection rectangle, a useful tool for doing several different things. It allows you to move things around the screen, delete parts of the machine (rather than the whole thing), and copy parts of the machine for use elsewhere, either in the same machine or in other machines.

Let's quickly try out a few of these things. We're going to have you make a copy of the text you just added, and move it to another place in the window. The first thing to do is click on the selection tool button. Then put the mouse cursor above and to the left of the text. Click down and drag the mouse until the selection rectangle completely surrounds the text, and then release the mouse button. This will leave a blinking rectangle around the text you have selected.

At this point, choose **Copy** from the **Edit** menu (or use Command-C). The text is now on the Macintosh's "clipboard," a temporary storage location that you can't see. Next, choose **Paste** from the **Edit** menu (Command-V). A rectangle containing a copy of your text now appears in the upper left corner of the window. To move it where you want, click down inside the rectangle and drag it to the desired location. To get rid of the blinking rectangle, click somewhere outside the rectangle.

Of course, you don't need two copies of this piece of text. So let's delete one of them, using the same tool. Select the text you want to delete by dragging the rectangle around it. Then hit the delete key on your keyboard. This will delete anything inside the selection rectangle.

You can also use these techniques to move, copy, or delete parts of a state diagram. We will explain more about this in Chapter 5.

3.4 Setting up a tape

It does little good to build a machine if you can't use it. We've already discussed how to set up a tape and run a machine in the last chapter, so we can be brief.

The machine you have built is designed for use on tapes that have ∗'s and blanks on them. (Though it will do something or other on any tape. What will happen if it is run on a tape that consists of ∗ ∗ ∗ ∗AA∗ ∗ ∗?)

Input: Assuming that the default alphabet is still selected, which it will be unless you have done something odd, all you have to do to put down a string of ∗'s is click on the squares where you want them to appear. If you click on a square that already has one, it will be erased. Now put a string of ∗'s on the tape. You can also type them directly on the tape using the keyboard, as explained earlier.

Position the Read Head: Move the read head where you want it, in this case to the leftmost ∗. You can do this by dragging it with the mouse or by using the left and right arrow keys.

3.5 Running the machine

Now you are ready to run the machine. Choose **Graphic Run** from the **Execute** menu.

Exercise 2 Try running the Eraser on several different tapes, starting in different positions: to the left of the string, in the middle of it, and to its right. Does the machine do what you expected?

3.6 Summary of tape

Sometimes a tape gets hard to read. Indeed, sometimes it is so long it doesn't all fit on the screen at once. Turing's World gives us a way to get a quick summary of the entire tape. Put down a lot of *'s, maybe leaving some blanks. Now pull down the **Display** menu and choose **Describe Tape** (or type Command-D). A window will appear that describes the entire contents of the tape, beginning and ending with an infinite number of blanks. (That is what the "inf" means: infinitely many.)

If you want a continuously updated tape summary while you are running a machine, select the item **Auto-Refresh** on the **Display** menu and then choose **Describe Tape**.

Exercise 3 Choose **Auto-Refresh** from the **Display** menu, and then **Describe Tape**. Then run the Eraser on a string of *'s, watching the tape description window.

3.7 Moving windows

The tape description window may have obscured the state diagram while you ran the machine. To move the window, put the cursor on the title-bar and drag the window where you want it. You can move the alphabet window and any state diagram windows in a similar fashion.

You can also move the tool palette window and the tape window. To move these windows, locate the cursor in the white margin at the edges of the window. When the cursor changes to crossed arrows, click and drag the window wherever you want it.

3.8 Doing things during a computation

By the way, requesting a tape description is one of several interesting things you can do in the middle of a computation (besides speeding it up or slowing it down). Another is checking to see how much time and space you've used so far, using the **Time-Space Count**, also on the **Display** menu. Try these out while running your machine on a long series of *'s.

3.9   Saving a state diagram

To save the state diagram of a machine, you choose either **Save State Diagram As...** or **Save State Diagram** from the **File** menu. There is only one case where there is a difference between these two commands. If you have been working on a machine that has already been saved before, **Save State Diagram** automatically replaces the old version with the new one, whereas **Save State Diagram As...** gives you a chance to name it something new, thus keeping the old version of the diagram, with the old name. This is useful when you want to borrow some other machine and modify it for some new purpose. It is always safer to use **Save State Diagram As....**

Tip: It is a good idea to save a state diagram when you first start building a new machine, so that it gets a name. The name is displayed at the top of the window, and from then on you can simply use Command-S to save the state diagram as you build it. It is always a good idea to save your work frequently, in case the power goes off or some other disaster strikes.

When you are asked for a name for your file, choose something suggestive, something that gives a hint of what the machine does. There is nothing more frustrating than trying to find a machine and not being able to tell which of ten machines on your disk it might be. (If you are turning in homework problems in a class, your instructor will tell you how to name them.)

You also want to make sure that you save your state diagram where you want it, in the right folder and on the right disk. At the top-right of the dialog box that appears when you choose **Save State Diagram As...**, you will see the name of the disk where the state diagram will be saved. This probably says Turing's World, if you are running the program off of the original Turing's World disk. To the left of the disk name there is a rectangular box that indicates where on the disk the state diagram will be saved. If, for example, the name Machines and Tapes appears in this box, then your file will be saved in the folder by that name. To move "up" the folder hierarchy, say to save the file at the top level of the Turing's World disk, you would click down on the rectangle containing the name Machines and Tapes and then choose Turing's World from the popup menu that appears. To move back down the folder hierarchy, say to get back into the folder Machines and Tapes, double-click on the folder's name, which now appears in the scrolling list.

Exercise 4 Save your ∗-eraser, naming it The Eraser, so that you can fool around with it but still have a saved copy on the disk. (We're going to use it later to learn how to edit state diagrams.) If you are new to the Macintosh, practice moving up and down through folders, as described in the last paragraph. See if you can save your file at the top level of your Turing's World disk, not inside any folder.

3.10 Changing the start state

Every Turing machine has to have a *start state*, a state for the machine to be in when the computation begins. This state is indicated on the state diagram by means of an arrow pointing to the state. Turing's World sets the default start state to be state 0. But it also gives you a way to specify a different start state.

Exercise 5 Build a machine that is just like the previous eraser, except that you start it on the blank square just to the left of the string of *'s you want erased.

The machine you have just built should look *very* much like our previous eraser machine. In fact, we could have used the previous machine and simply changed the start state to state 1. Open the Eraser that you have saved on your disk. Using the node tool, click once on the node that you want to become the start state, in this case, node 1. Then choose **Set Start State...** from the **Execute** menu. You will see that the arrow indicating the start state now points at node 1. Try out the machine on a string of *'s, making sure to locate the read head on the blank square to the left of the string.

3.11   Stopping a runaway Turing machine

Sometimes you will want to stop a machine in the middle of a computation. Maybe you want to alter the tape or the machine, or get something to eat and don't want to miss the action.

You can temporarily stop a machine in the middle of a computation by choosing **Pause** from the **Execute** menu (or type Command-P. You can resume the run, if and when you want, by typing Command-G, or choosing any of the run commands.

You can call the computation to a permanent halt by selecting **Reset Machine** from the **Execute** menu (or by typing Command-M). You can either do this after pausing or while the machine is in operation. When you reset the machine, you can then change it or the tape any way you want. If you just pause the computation, you will not be able to change the machine or the tape, since you are still in the middle of a computation.

Exercise 6 Modify the Eraser machine by adding an arc from node 0 to node 0 labeled $(-, \Rightarrow)$. To draw an arc from a node to itself, just click on that node twice using one of the arc tools. Go ahead and run this machine.

Now you see yet another reason for needing to stop a machine in mid-computation. Our new machine would never halt on its own. You will have to reset this machine or else it will run forever.

3.12 Using a custom alphabet

To design a machine using a larger alphabet, choose **Custom Alphabet...** from the **Alphabets** menu. (You can even change your mind about the alphabet in the middle of designing a machine.) Suppose you want to use an alphabet consisting of *, A, B, and for some reason you want * to be the symbol that initially fills all the squares on the tape (a role usually reserved for the blank symbol).

Choose **Custom Alphabet.** Click on the boxes next to A and B. An X will appear in the boxes. Chances are * already has an X next to it, which is what we want. However, we do not want anything else in our alphabet, so if there is anything else with an X next to it (like blank), get rid of it by clicking its X. Now click OK.

At this point you will be asked for a choice of characters to fill the tape with, so select * and then click OK. You will now need to refresh the tape by choosing **Clear Tape** from the **Execute** menu.

Exercise 7 Having created this alphabet, put AA*BB on the tape. Predict what the tape summary will show. Check your answer.

Exercise 8 Design a machine, using the above alphabet, that starts at the left of a string of A's and B's (on a tape otherwise filled with *'s), and replaces all the A's with *'s.

To get your tape back to normal, choose **Default**

Alphabet from the **Alphabets** menu. Then choose **Clear Tape** from the **Execute** menu.

Coding with the default alphabet

One of the many things that Turing showed in his original paper was that any symbolic computation could be simulated by one that uses only two symbols, say blank and ∗. This is very surprising at first glance, but it is also very important. It allows us to store information with the basic "on" and "off" that electricity provides and that digital computers run on.

Turing's idea was simple enough. Just think of it as a problem in cryptography. Find a way to encrypt words written in a large alphabet into one with just two symbols. For example, how would you encrypt A, B, − using just ∗, −? One natural way that works out pretty well in practice is to use the following code:

−	encrypted by	− −
A	encrypted by	∗ −
B	encrypted by	− ∗

Thus, for example, the string ABBA−AB would be encrypted by ∗ − − ∗ − ∗ ∗ − − − ∗ − −∗.

This result explains why so much theoretical work assumes that one is dealing with an alphabet containing just two symbols. But in the actual design of Turing machines, restricting to two symbols is impractical. For this reason, Turing's World lets you build machines using quite a few symbols. In Exercises 17–20, Chapter 8, we will explore techniques for encrypting larger alphabets using the default alphabet.

4

Editing a State Diagram

A pacing machine

To teach you about editing, we are going to build a little Turing machine that simply moves back and forth between the two ends of any finite string of *'s. It is pretty obvious that any such machine has to have at least two states, one where it is moving left looking for the left end of the string, and one where it is moving right looking for the right end. Let's call these states 0 and 1. If the machine is in state 0 and sees a *, then it stays in state 0 and moves right. However, if it is in state 0 and sees a blank, then it moves left and goes into state 1. Conversely with state 1. A picture of this machine can be found in Figure 4.

Rather than start from scratch, let's build this machine by editing the diagram of the Eraser you built in the last chapter. This will give us a chance to explain many of the editing features of Turing's World. So open the state diagram of the Eraser. It should look like Figure 3 on page 45. Our aim is to transform this diagram into the state diagram shown in Figure 4.

4.1 Editing text

The first thing you should do is modify or delete the comment describing what the machine does. You can do either by selecting the text tool (**T**) from the tool palette, and then clicking on the piece of text in ques-

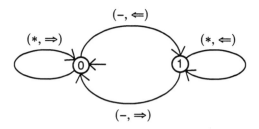

Figure 4 The Pacer

tion. The text will become highlighted. Once it is high-lighted, you can delete the whole piece of text by hit-ting the delete (backspace) key on the keyboard, or you can replace it with other text by just typing the new text.

If you want to insert some new text in the middle of the existing text, then after it is highlighted, click where you want the new text. The vertical line that appears is called the *insertion point*; whatever you type will now appear there.

If you want to delete some portion of the text, but not all of it, then put the insertion point at the be-ginning of the text to be deleted and drag it until the highlighting covers the part to be deleted. Again, you can either delete the highlighted text (with delete), or replace it by typing new text.

Experiment a bit with these editing techniques. In particular, try inserting new text at various points, in-cluding the beginning and end of a line.

When you are finished editing a piece of text, click elsewhere in the window to indicate that you are fin-ished.

Exercise 1 Try out these procedures by replacing the original caption with Paces up and down a sequence of *'s. Make a few mistakes and correct them.

4.2 Moving text

You can also move text around. Choose the selection tool from the tool palette. Drag a rectangle around the piece of text you want to move, and then release the mouse button. Place the mouse cursor inside the blinking rectangle and drag it to the desired location. Try this out by centering your description of the Pacer.

Notice that you can only move, cut, or copy a whole piece of text using the selection rectangle.

There is another way to move text that is sometimes faster than using the selection rectangle. If you choose the text tool and move the mouse until the X cursor appears over the text you want to move, then you can click down and drag the text wherever you want it. Don't click and release the mouse button, though, since Turing's World will think you want to edit the text: click and immediately drag the text.

4.3 Moving and editing arcs

Next we want to edit the top arc. (The bottom one is ok as it stands.) To move or edit an arc, select one of the arc buttons from the tool palette. You can now drag an arc around by its label. (If you ever end up with arcs that are on top of each other, or are hard to read for some other reason, you can use this feature to move one or both of them somewhere else.) You can delete an arc by clicking on the label and then typing delete after the label starts blinking. (Draw an extra arc and delete it just for practice.)

One way to change the label on an arc is to delete the arc and then redraw it. But there are better ways to edit the label. You can either (a) double click on the label, or (b) click once on it and then choose **Edit Arc** from the **Edit** menu. This leaves the arc in place and lets you give it a new label in the old way.

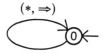

Figure 5 The One Way Pacer

Exercise 2 To finish the pacer, edit the label on the top arc and then add the two remaining arcs. Your pacer is now complete. Save it and then test it to see if it works.

4.4 Moving and editing nodes

A one way pacing machine

Suppose we want to modify the Pacer so that it only goes one way, say, right, and then halts. All we need to do is delete state 1 and all the arcs connected to it. This would give us the diagram shown in Figure 5.

To edit a node, select the node button on the tool palette. With the node tool selected, you can move nodes around by clicking and dragging them, releasing the mouse button when they are positioned where you want them to be. The arcs attached to a node will readjust to the node's new location.

To delete a node with the node tool, simply click on it and then type delete (backspace). Try deleting node 1. Notice that deleting the node automatically deletes all arcs connected to it.

4.5 Editing with the selection rectangle

You can also move portions of a state diagram around using the selection rectangle. Suppose you want to move a group of several nodes, along with all arcs attached to them. To do this, drag a selection rectangle around the nodes in question and then drag the rectangle to the desired location. (You may have to rearrange things a bit first, to make the desired nodes fit in a rectangle that does not contain any other nodes.) Try this out when you build the machine in the following exercise.

Exercise 3 Open the Lone Rearranger and change the alphabet so that it includes the symbol C. Alter this machine so that it alphabetizes strings of A's, B's, and C's. Before making any changes, though, save the diagram as ABC Rearranger. Then you won't accidentally write over the Lone Rearranger.

5

Using Submachines

One of the most useful features of Turing's World for building complicated machines is the ability to package a machine into a single node to be used as a submachine in other machines—machines which can themselves be submachines of other machines, and so on.

The ability to use submachines does not increase the power of Turing machines in theory, but they do increase our ability to build comprehensible yet sophisticated machines in a reasonable amount of time. To emphasize the fact that they don't allow us to do anything that couldn't be done without them, our first use of them will be to give a different diagram of the Pacer, one that uses submachines.

5.1 Expanding a node into a submachine

Any node can be changed into a submachine, simply by double-clicking on the node. A new window then appears, and you can draw the submachine's state diagram in the ordinary way. When you are finished, close the window and it will shrink back to its previous location, but now in the form of a square.

Exercise 1 Start a new machine using the default alphabet. Drop two nodes for a pacer and connect them with the same arcs used in the earlier pacer. But don't draw the loops (i.e., the arc from node 0 to itself and the arc from node 1 to itself). Select the node button and double-click on node 0. Then draw a submachine with a single node with an arc to itself, labeled $(*, \Rightarrow)$. Close the window and build a similar submachine in node 1. The arc in this submachine should be labeled $(*, \Leftarrow)$. Close the window and test your new pacer. (You'll have to stop it by resetting the machine.)

The way a submachine works is this. When the machine enters a state which is in fact a submachine, it enters the starting state of that submachine. The computation then proceeds as usual in the submachine until we get to a state where the submachine would ordinarily halt, that is, a state where there are no instructions involving the currently scanned symbol. Then the control is passed back to the main machine. If there are instructions at the square node for the scanned symbol, then the machine continues on, if not, everything comes to a halt.

Note that you can nest submachines as much as you like. In other words, you can have submachines with further submachines, and so forth. All the procedures for building sub-submachines are the same as for building the original submachines.

5.2 Moving and sizing submachine windows

When you first double-click on a node to open a submachine, the window that appears is fairly small. If you'd like to change its size, drag the sizing box in the lower right corner of the window. To move the window, drag it by the title-bar. The new size and location will be remembered whenever the submachine window opens up. Try this out on your new pacer.

5.3 Shrinking the main machine

Besides changing a node of the main machine into a submachine, you can also shrink the main machine into a single submachine. Just for fun, try this out on your new pacer. Make sure both of its submachines are closed and choose **Main** → **Submachine** from the **Edit** menu. Your machine will be packaged up into a little box and given the label 0. Expand the new node 0 (double-click) and see what you find. You can reverse this process by choosing **Submachine** → **Main** from the **Edit** menu. This command turns the open submachine into the main machine, but only works if there is only one submachine in the state diagram.

5.4   Cutting or copying a submachine

When building complicated machines, you will often want to reuse parts of your machine or some other machine. Turing's World allows you to cut or copy any portion of a state diagram to the Macintosh clipboard, and then to paste it into a machine wherever you want it.

To try this out, let's build a machine that does the following. It takes as input a sequence of *'s, with at most one blank in the middle of the sequence. If there is *no* blank, then it returns to the extreme left of the sequence and stops; if there *is* a blank, then it continues on and halts at the extreme right. Figure 6 contains a diagram of the machine we'll construct; node 0 and node 3 are both submachines that move right along a string of *'s, while node 2 is a submachine that moves left and stops.

The first thing we want this machine to do is move right as long as it sees *'s. We already have a submachine that does this in the new pacer, so for practice copying let's just reuse it. The submachine we want is contained in node 0. To copy it, drag the selection rectangle around node 0, being careful not to capture node 1. Your rectangle will have to include pieces of the arcs that connect node 0 and 1, but that won't matter. Once

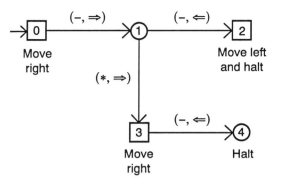

Figure 6 Blank Check

the rectangle is in place, choose **Copy** (Command-C) from the **Edit** menu.

Both **Cut** (Command-X) and **Copy** (Command-C) leave a copy of the selected material on the clipboard so that you can paste it elsewhere as often as you want, either into the machine now displayed or into another machine. The difference is that **Cut** removes the material from the original, while **Copy** leaves it in place.

We want to paste what we have on the clipboard into a new machine, so get a new state diagram by choosing **New State Diagram** from the **File** menu. (You will be asked whether you want to save changes to your current state diagram. There is no need to.) Once the new window appears, select **Paste** (Command-V) from the **Edit** menu and the material on the clipboard will appear in the upper left corner of the window. Move it wherever you want. Click outside of the rectangle to "install" the pasted material in the window.

Notice what happened to the arcs coming into node 0 from outside the selected rectangle. When you paste into the new window, it looks as though a portion of these arcs is still around. But when you install the node, they disappear, since they have lost their connections.

To finish the machine, drop nodes 1 and 2 and connect them with arcs as in the above figure. For node 3 we want another copy of the submachine stored on the clipboard, so choose **Paste** again and a new copy will appear. Move the node where you want it, and click outside the rectangle to install it. Notice that the node magically gets renumbered with the first available number not in use, in this case, 3.

Exercise 2 You should be able to finish this machine on your own. You will want to double-click on node 2 and design a submachine that returns left to the starting position, and also add node 4 (so the machine halts on the rightmost * rather than the blank next to it).

You no doubt noticed that the clipboard keeps the copied material so that it can be reused repeatedly. But if you cut or copy something else, it will replace what is on the clipboard with the new material.

It would be a good idea for you to practice copying and pasting portions of your machine. Check to see what happens if you copy several nodes and paste them back into the same machine. See what happens to an arc if it is partially outside the selection rectangle but its source and destination are both inside the rectangle. What happens if you select both text and some nodes?

You can paste material from Turing's World into other kinds of documents, such as word processing files. This is very useful if you have designed a complicated machine and are writing a report on how it works.

5.5   Editing a submachine

You can edit a submachine using exactly the same procedures as those used in editing the main machine. It often saves time in building complex machines to copy a submachine (or any other portion of your state diagram) that is close to what you want, and then edit it so that it works the way you want it to.

Exercise 3 Delete node 4 in the machine you just built and edit submachine 3 so that the new machine behaves just like the old one.

5.6   Doing too much with a submachine

Here is a hint for constructing machines with submachines. Remember that there is no way to have a transition from one node to another that does nothing at all. You always have to print, or move left, or move right. This can be a bother when you first start building machines by hooking together submachines, since you will be tempted to have the submachine do one more thing than it should. You need to save some action for the large machine to carry out in getting to the next node.

Exercise 4 (Symbolic Commutative Algebra) We are now ready to build the machine we promised you at the beginning of Chapter 2. However, rather than do it for you, we are going to let you do it yourself, to see if you have mastered all this. Use three of the prepackaged Turing machines (Paren Check, Paren Remover, and Lone Rearranger) to build a machine which combines features of all three. Your machine should start with an arbitrary string of A's, B's, ('s, and)'s, and see if it is properly matched. If not, it should do just what the parenthesis checker did: print BAD. However, if it is properly matched, it should remove all parentheses and then reorder the result so that all A's come before all B's. You will want to use all three of the prepackaged machines as submachines, but you will need to modify at least one of them, the parenthesis checker.

Hint: Do things in the following order. Create a new state diagram called CommAlg. Save this. Open the parenthesis checker, turn it into a submachine, and cut it out. Close the parenthesis checker without saving the changes. Now reopen CommAlg. Paste the copy of the parenthesis checker currently on the clipboard into this diagram. It will be labeled 0. Save CommAlg again, and then get the other two machines in the same way, in turn. Once they are all in CommAlg you can start hooking them together. For example, the first thing you will want to do is open the parenthesis checker and delete the submachine that finishes up and writes GOOD. Then you need to hook the submachines together.

6

Other Features of Turing's World

6.1 Debugging a Turing machine

Turing's World makes debugging Turing machines a joy. Well, if not a joy, at least a lot easier than ever before, since it allows the design and testing phases to interact. The best way to build a machine is not to build it and then debug it, but to build it up bit by bit, testing it as you go, using the various features provided by Turing's World. The main way to do this is by judicious use of the **Pause**, **Step**, and **Reverse** commands on the **Execute** menu.

We have already discussed **Pause** and **Step**, but we haven't told you yet how **Reverse** works. Notice that backing up through a computation is not like going forward, since there are often many different situations which could have gotten the machine and tape into the state they are in. So there is no way to look at a machine and tape and tell how you got there, in general.

To get around this problem, when Turing's World is running in graphic or step mode, it keeps track of the 20 most recent steps in the computation. You can then use **Reverse** (Command-R) to step backwards one step in the computation, and this can be repeated 19 times. Going back and forth through a tricky part of a computation can often show you what is wrong with your design.

Sometimes in debugging a machine, you will want to start it in some state other than the start state, say if you want to test a new part of your machine without running an entire computation. This can be accomplished by using **Set Start State...** from the **Execute** menu. Be sure to change it back when you are finished testing.

Exercise 1 Design a machine like our parenthesis remover, but one that is much more efficient. Test it as you go.

Tip: When you are building or debugging a Turing machine, always make sure you test the exceptional cases: input tapes that are not like the typical ones you had in mind in designing the machine. Design errors often show up on inputs like the empty sequence or sequences where only one symbol appears. The most common mistakes come from not considering these atypical inputs.

6.2   Using the wild card in arc labels

There is one very useful feature of Turing's World which we haven't illustrated at all in our prepackaged machines. This is the use of "wild cards" in building arcs.

When you label an arc, we've had you stay away from the option labeled otherwise in an elipse to the right of the letters. If you choose this, then the arc will be labeled with a pair that starts with "...", for example, $(\ldots, \Rightarrow)$. What this means is that if the machine sees any symbol for which no other action is specified, then the machine should move right.

The use of the wild card can greatly simplify the building of machines, especially when the alphabet has more than two or three symbols. For example, we could have simplified our parenthesis checker quite a bit if we had used it.

When you have a fixed alphabet, it's clear that any machine written using the wild card could be written without it. But when you allow the alphabet to vary, the machines have a different character with the wild card option present.

Exercise 2 Using the wild card option, build a machine that will simply move right forever, no matter what string in what alphabet it is given. Show that there is no Turing machine without the wild card that has this property.

Exercise 3 Go back and simplify Paren Check using the wild card option. Look for submachines that look like daisies: that's where it is most useful.

Exercise 4 Using the wild card feature, build a base-ten adder, that is, a machine which takes arbitrary pairs of arabic numerals and replaces them with their base-ten sum. For example, given the pair of numerals 978 and 35 separated by a blank, the machine should end up with the numeral 1013. Design you machine so it is started with the read head on the blank between the numerals and so it ends immediately to the right of the answer.

6.2.1 Getting into and out of submachines

In Chapter 5, we observed that there is sometimes a problem in getting smoothly out of a submachine. This problem can be solved by using the wild card feature, since it gives us a simple way to simulate the command: in state n, no matter what you see, leave it alone and go into state m.

To do this, introduce a new node k, an arc from n to k labeled $(\ldots, \Rightarrow)$ and an arc from k to m labeled $(\ldots, \Leftarrow)$. The sum total of these two is simply to get the machine from n to m. These are especially useful when one or both of n and m are submachines.

6.3   Schematic machines

Submachines provide a convenient way to build *schematic machines*, Turing machines which have one or more "holes" into which we can plug other machines. To build a schematic machine, simply build a machine with a submachine node containing nothing at all (except text, which keeps it from turning back into an ordinary, round node when it's closed). You can then paste any other machine into the empty submachine: just copy the whole machine and paste it into the submachine using the selection tool. What used to be the empty submachine will now behave exactly like the original machine. This is a very important technique, since it quickly allows you to assemble ways of building quite sophisticated Turing machines.

Suppose, for example, that we wanted a general way of taking a machine M and forming a new machine, call it M^2, which would first run M and then, if and when M halted, run M again. (Thus if M changed all A's to B's and vice versa, then M^2 would do this twice. Of course this would have the effect of putting you back where you started. See Exercise 7.) We can create a schematic machine for M^2 as follows. On a new state diagram, drop node 0 and double-click on it to turn it into a submachine. In this window write: This is the place to paste any machine you want to iterate. Paste the same machine in submachine slot 1. Then close the submachine. Now **Copy** and **Paste** this submachine to get submachine 1. Open it and make the obvious modification of the text. Finally, use the technique from the last section to get an arc that takes you from submachine 0 to submachine 1 without affecting the tape.

Exercise 5 (Iterater) Build the machine just described

and save it as Iterater. Make sure you annotate the machine so that it is clear what to do with it.

Exercise 6 (**Reverse video**) Build a Turing machine that will take a finite sequence of A's and B's and replace each B with A and vice versa. Have the machine start and end on the leftmost symbol in the string.

Exercise 7 Paste the reverse video machine from Exercise 6 into the Iterater, and see if it has the expected effect. Make sure you save this state diagram using some name other than Iterater, so that the Iterator does not get messed up.

6.4 Using 4-tuples to describe machines

You should not make the mistake of confusing the state diagram of a Turing machine with the Turing machine itself, even though in this book we haven't been careful to distinguish them. But confusing the two is like confusing the blueprint of a house with the house itself: a good way to get wet. State diagrams are simply a convenient way to *represent* Turing machines. There are many others, both in theory, and in use.

One way to represent a Turing machine is by means of a set of "4-tuples." Each such 4-tuple represents the same information as is represented by one arc of a state diagram. Open your original Pacer (the one without submachines) and then choose **Show Tuples** from the **Display** menu (or type Command-U). You will see a new window appear, showing the following set of 4-tuples:

$$0 * \Rightarrow 0$$
$$0 - \Leftarrow 1$$
$$1 * \Leftarrow 1$$
$$1 - \Rightarrow 0$$

You can think of this as a compact notation for a

very primitive programming language. You should read the first two instructions, the ones with a 0 in the first place, as: "If you see a *, then move right and go to 0; if you see a blank, move left and go to 1."

If your machine contains submachines, the numbers corresponding to these nodes will appear in the 4-tuple list as decimals. For example, if state 3 is a submachine whose start state is 1, then this will appear in the 4-tuple list as 3.1.

Representing Turing machines by means of 4-tuples is important for the general theory, because it allows us to represent any Turing machine M by means of a string of symbols α_M. And strings of symbols are the sorts of things Turing machines themselves operate on. This gives rise to the possibility of having a single "universal" Turing machine that can mimic any Turing machine, a possibility Turing showed is in fact realizable. This is discussed further in Chapter 8, page 106.

Exercise 8 Step through the Pacer with the 4-tuples displayed. What does the highlighting of the 4-tuples represent?

Different formalizations of the notion of a Turing machine define them in slightly different ways. The main difference is in how much a machine can do in a single step: Does the machine have to choose between printing a symbol on the tape and moving along the tape, or can it do both in one step? The former is called the "4-tuple" approach, the latter the "5-tuple" approach, for obvious reasons. There is no theoretical difference between the two approaches. (See Exercise 33, page 98.)

Tip: You can display the 4-tuples while building a machine and the list will be actively updated. If you've let the diagram get too messy, the 4-tuple window can sometimes be a help in figuring out where arcs go and how they are labeled.

6.5 Making room for large state diagrams

We've already discussed the fact that there is more room for your machines than fits on the screen at one time. You can make room to the right by sliding the bottom scroll box to the right. To make more room below, choose **Add Page** from the **Edit** menu, and then move the vertical scroll box.

Tip: Sometimes you will want to draw an arc between nodes that do not seem to fit on the screen at one time. Remember that you can always move nodes closer together (using the edit node feature), draw your arc, and then move the nodes back to their desired position.

6.6 Printing

If you are using Turing's World on your own, or to help you do homework in a course where it's not required, you may want to print out the state diagram of a Turing machine. Or you may want to print out the state diagram of a particularly interesting machine to send to the authors, who would be delighted to receive such. This is easy enough, using the **Print** commands on the **File** menu.

The three **Print...** commands on the **File** menu are self-explanatory. They allow you to print a state diagram of a machine, the 4-tuple description of a machine, or the summary of a particular tape, respectively. If you print the state diagram of a machine that contains submachines, each submachine will appear on a separate page.

7

Other kinds of machines

In the Chapter 1, we mentioned variations on Turing machines that have become important in computer science. Turing's World gives you the capacity to design some of these other kinds of machines, in addition to ordinary, deterministic Turing machines.

7.1 Finite state machines

The main difference between a Turing machine and a finite state machine is that finite state machines cannot write on their tape. What this means is that the tape cannot be used as working memory as the machine runs through a computation. For example, in the parenthesis checker machine, the space to the left of the initial string is used to "remember" how many unmatched left parentheses the machine has seen so far. It can remember arbitrarily many, since the tape is infinitely long.

A finite state machine, by contrast, can only "remember" something by being in a state that represents that particular fact. Since a finite state machine can only have a finite number of states, its memory is finite. This is why these machines are called finite state machines, to contrast the finiteness of their memory with the unlimited working memory of Turing machines.

By the way, the task of matching parentheses is one that cannot be carried out by any finite state machine, no matter how large. We will make this plausible in an exercise. To actually prove this fact, one uses something known as a "pumping lemma," which you may learn about in class.

7.1.1 One-way finite state machines

A one-way finite state machine is simply a Turing machine that has only one possible action: move right. If you choose **One-Way Finite State** from the **Machine** menu, then when you create arcs, you will only be asked for the input character; the right arrow ($\Rightarrow$) will be filled in automatically. (If the current state diagram has arcs indicating actions other than moving right, you will not be able to change the machine type to one-way finite state.)

The way to think about such a finite state machine is that it is being fed an input which it scans one character at a time, left to right, keeping track of various features by means of its internal states. It comes to a halt when it reaches the right end of the input string.

7.1.2 Acceptance nodes

You might wonder what good a machine like this could possibly be, if has no way to produce any output. To get around this, finite state machines have one new feature not possessed by Turing machines. Some of their states, called *acceptance states*, are considered special. If at the end of reading the input, the machine is in one of the acceptance states, the machine "accepts" the input; otherwise it rejects the input.

Let's build a simple finite state machine and test it. The machine is shown in Figure 7. This machine is designed to test a string of A's and B's, to see if it needs to be shipped off to the Lone Rearranger. In other words, it checks them to see whether there are any A's that follow B's. If there are, then it rejects the string. Otherwise, it accepts the string.

To build this machine, first choose **One-way finite**

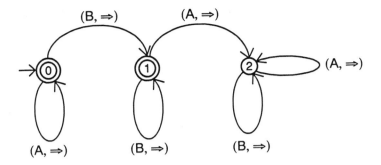

Figure 7 The Inspector

state from the **Machine** menu. Then specify your alphabet to consist of just A, B, and blank. Now build the machine shown in the figure, using regular nodes. When you are finished, choose the acceptance node tool, the second tool on the palette. Click once on each of nodes 0 and 1, to turn them into acceptance nodes. If you make a mistake and click on 2, just click it again to change it back to an ordinary node.

Exercise 1 Run your Inspector on a variety of different strings, to see how it works. Don't forget to try it on a completely blank tape, and on tapes consisting of all A's or all B's. Does it do the right thing in these cases?

With finite state machines, it is assumed that the input must be completely traversed in order for the string to be accepted. Machines that stop midway through the input do not count as accepting the string, even if they happen to stop in an acceptance state.

Exercise 2 Change the alphabet of the Inspector to include C, and then run it on the string AABBBC. Does it accept or reject this string? Why?

One final point. Turing's World does not allow you to have accepting states inside submachines. They must be at the top level of your machine. If you try to change a node inside a submachine into an accepting node, your Macintosh will beep at you.

7.1.3   Two-way finite state machines

Two-way finite state machines are just like the one-way variety, except they can move either left or right. Unlike their simple one-way cousins, these machines can get into an infinite loop. If they do, the string is not accepted, so it is considered rejected.

You might think that the ability to look back and forth on a tape would give you a lot more power. But it turns out that any two-way finite state machine can be mimicked by a one-way finite state machine, though sometimes at the cost of building a bigger machine. We will give an example of this in the exercises.

To build a two-way finite state machine, choose **Two-way Finite State** from the **Machine** menu. When you build an arc, you now have to specify the action, however your choices are limited to moving right and moving left.

Exercise 3 Build a two-way finite state machine with the following 4-tuple description. State 0 is the start state and state 1 is the only accepting state.

$$0 \ A \Rightarrow 0$$
$$0 \ B \Rightarrow 0$$
$$0 \ C \Leftarrow 1$$
$$1 \ B \Leftarrow 1$$
$$1 \ A \Rightarrow 0$$

Describe the set of strings accepted by this machine. Remember that if the machine does not halt, then the string is not accepted.

7.2   Nondeterministic machines

In all the machines that we have discussed so far, there is a basic requirement that if the machine is in a given

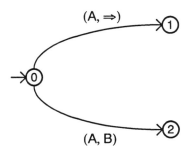

Figure 8 A nondeterministic split

state, scanning a given symbol, there is at most one action that the machine can carry out. Thus if you are building an ordinary Turing machine and you try to label an arc with an input character that is already used at that node, Turing's World warns you of that fact. If you go ahead and put down such an arc and then run your machine, Turing's World simply ignores the first arc.

What if we allowed machines that could have multiple arcs labeled with the same input character emerging from a single node? We could think of this as a way of specifying a "parallel" computation, one that splits and goes along different paths when it gets to such nodes. Such machines are called "nondeterministic" machines.

Suppose, for example, that we had a machine whose state diagram contained the nodes and arcs shown in Figure 8. Notice that if the machine is in state 0 and sees an A, it is given a choice of actions. It can either move right, going into state 1, or print a B and go into state 2. The way a nondeterministic machine works is that it does *both*, in parallel, hoping for one of the two paths to come to a successful conclusion. If so, the other path is abandoned.

Turing's World allows you to build nondeterministic versions of each of the machines we have discussed:

one-way finite state, two-way finite state, and Turing machines. To build a nondeterministic machine, you simply choose **Nondeterministic** from the **Machine** menu. You can choose this at any point during the construction of the machine.

7.2.1 Nondeterministic finite state machines

Let's look first at a nondeterministic finite state machine. Open the machine called 4 A's or 3 B's. This machine accepts those strings of A's and B's that contain at least four in a row of the former or three in a row of the latter.

Having opened the machine, look at the **Machine** menu. You will see that **Nondeterministic** has a check next to it for the first time. This is because the machine was built and saved as a nondeterministic machine.

To get a feel for the way these machines work, open Tape 5. Look at this tape and notice that it does indeed contain a sequence of four A's, but fairly far down the line. Run the machine on Tape 5 at graphic speed. As the machine runs, you will see multiple tapes appear, one for each computation. Since we do not have enough screen space to duplicate the state diagram for each computation, you will instead see more than one highlighted node, corresponding to different computations. Each tape has a small number at the left telling you which state that computation is in.

As the machine runs, you will hear an occasional beep and, simultaneously, see a tape disappear from the screen. This indicates that the corresponding computation has come to an unsuccessful conclusion. The computation has halted in a non-accepting state, that is, it has either come to the end of the string and is not in an accepting state, or it has halted before reaching the end of the input string. But on this particular tape, the machine eventually finds a successful computation, and so accepts the string. Once the string is accepted, any other computations that are still running are abandoned as well, leaving only the tape on which the successful computation was run.

Exercise 4 Run the machine on several different tapes, some that contain both four A's and three B's, some that contain neither, and some that contain one but not the other. Try to get a good feel for what is going on. Try running them at various speeds.

7.2.2 The process tree

Nondeterministic computations can be pretty hard to follow, in fact down-right confusing, due to the multiple splitting that takes place. Turing's World provides a graphic depiction of this splitting process that may help you picture what is going on.

Open 4 A's or 3 B's again, if it is not already open. If you have recently run the machine, choose **Reset Machine** from the **Execute** menu. Create a tape with the string AABBB on it and with the read head on the first A. Before running the machine, choose **Process Tree** from the **Display** menu. Resize the process tree window, making it about three inches tall. Then run the machine at graphic speed.

As the machine runs, you will see a splitting tree that depicts the overall state of all the computations. The numbers at the ends of the branches indicate the current state of that computation. The numbers in italic correspond to computations that have "died off." The numbers in bold indicate the computations that are still running and whose tapes are displayed. At the end of your run, all but one of the tapes will go away, and the corresponding numbers in the process tree will revert to regular, non-bold type.

If you want to see any of the tapes corresponding to these other computations, simply click on the numbers corresponding to them. If you hold down the shift key,

you can display several at once. (The tapes corresponding to the dead computations, those with italic numbers,
cannot be displayed. They've gone to the great tape library in the sky.)

At any point in a computation, you can pause the
machine, activate the process tree window, and click on
just the numbers whose tapes you want to display. This
does not end the other computations, but it does hide
their tapes. This allows you to focus on one or more
paths in the computation.

Exercise 5 Modify the 4 A's or 3 B's machine by deleting nodes 5, 6, and 7. Run this machine on a string
of A's and B's, with the process window open. Notice
that the machine is still nondeterministic, and is still
splitting. What strings does it accept? How would the
machine behave if you made it deterministic by deleting
the arc from node 0 to itself labeled A?

7.2.3 Nondeterministic Turing machines

Turing's World also allows you to add nondeterminism
to Turing machines. But this addition requires a slight
change to the way we handle the end of a computation. To emphasize this, we observe that if you take
the Lone Rearranger and run it as a nondeterministic
Turing machine, you will be told "No successful computation found" at the end of the computation.

When you create nondeterministic Turing machines,
you need to specify which states count as successful
stopping points, or halting states. This is done by using
the acceptance node tool from the tool palette. Thus, in
this section we will refer to any node that has a double
circle around it as a *halting node.*

In a nondeterministic Turing machine, if a computation ends in a non-halting state, then it is assumed
to be unsuccessful, so that the remaining computations
should continue. Thus, the halting states are what tell

the nondeterministic machine when it can stop computing.

Exercise 6 Open the Lone Rearranger and Tape 1. Choose **Nondeterministic** from the **Machine** menu. Run the machine. Now change node 1 to a halting node with the acceptance tool, and rerun the machine on a new copy of Tape 1.

Other than specifying the halting states, you build nondeterministic Turing machines in the same way as deterministic Turing machines. Of course nondeterministic machines can have multiple arcs with the same input character coming from the same node.

Exercise 7 Design a nondeterministic Turing machine with three states that starts at the left of a string of A's and B's, and ends with either a string of all A's or all B's, depending on whether or not there are more A's than B's in the original string. In other words, the successful computation should be the one that has to change the fewest symbols. Test your machine on various tapes, until you are confident that it is running properly. What does it do when there are equal numbers of A's and B's? Can you explain why? [Hint: When a computation splits, it first proceeds down the last arc drawn.]

8

Additional Exercises and Projects

The authors would enjoy getting copies of particularly elegant solutions to any of these exercises, or other interesting machines the reader may build. Please send disks or hardcopy to either of us. We will be happy to return your disk if so requested.

8.1 Finite state machines

The problems in this section assume that you have read about finite state machines in Chapter 7. You can skip this section if you want to move directly to Turing machines.

Exercise 1 (Dragonslayer) Imagine a knight on a journey through a land infested with dragons (D's) and evil trolls (T's), as well as some friendly civilians (F's). He starts his journey with no weapons, but along the way he can find swords (→'s), used to slay dragons, and acquire spells (#'s), used to enchant the stupid trolls. He can carry at most one of each at any one time, and they are gone once he uses them. Luckily, people seem to have dropped a lot of swords and spells around the landscape.

Design a one-way finite state machine that accepts a string if and only if the string represents a journey on which the knight survives. Thus, for example, it should accept F→FD#→TFD but it should reject F→FD#→TFDT, since the final troll gets the best of him.

Exercise 2 (Lights out) Imagine you have three light switches, numbered 1, 2, and 3, controlling two lights in a hallway, one at the north end, the other at the south end. The way the switches work is as follows:

1. Switch 1 turns the north light on or off, depending on whether it is currently off or on.
2. Switch 2 switches both of the lights, changing them from on to off or off to on.
3. Switch 3 turns the south light on or off.

Design a one-way finite state machine that checks to see if both the lights are off. That is, it should accept a string of 1's, 2's, and 3's if and only if the string represents a sequence of switch flips that results in both the lights being off (assuming they were off to start with). For example, it should accept the string 213 since this turns on both the lights and then turns off the north light and then the south light. By contrast, it should reject the string 23213 since this sequence of flips leaves the north light on.

Exercise 3 (0-1-2 Cube) Design a one-way finite state machine that accepts a string of 0's, 1's, and 2's if and only if the string contains an odd number of each numeral. Thus it should accept the string 1121022, but not 1120022.

Exercise 4 (Casting 3's) A well-known trick for seeing if a number expressed in base-ten notation is divisible by three is to add up the digits and see if the sum is divisible by 3. Using this trick, design a finite state machine that runs through a base-ten numeral and accepts it if and only if it represents a number divisible by 3. You should be able to do this with a machine having just three states.

Regular languages and finite state machines

A finite state machine using a given alphabet is thought of as determining a set of strings, the strings made up from letters in that alphabet that are accepted by the machine. This is called the *language* accepted by the machine. It is often denoted by $L(M)$, where M is the machine in question. A set of strings is said to be a *regular language* if it is the language accepted by some finite state machine. The theory of regular languages is a central topic in computer science.

Exercise 5 (Regular languages) Build one-way, deterministic finite state machines which show that the following are regular languages. Assume that the alphabet contains just A and B.

1. The set of strings that end in BBB.
2. The set of strings that do not end in BBB.
3. The set of strings that contain at least five consecutive B's. Thus it would accept BABBBBBAB but it would reject BABBBBAB.
4. The set of strings that do not contain at least five consecutive B's.

Exercise 6 (Equal A's and B's) Design a one-way deterministic finite state machine that operates on the set of strings of A's and B's, and accepts just those strings that contain the same number of A's and B's, and have the additional property that no prefix contains more than one more of one numeral than of the other. Thus, for example, it should accept BAABAB and ABABBA but it should reject ABBBAA since the prefix ABBB contains two more B's than A's. [This exercise is suggested by Exercise 2.7 in [Hopcroft & Ullman]. It is interesting because the set of strings that contain the same number of A's and B's is not a regular language. In other words, there is no finite state machine that accepts all and only these strings.]

Exercise 7 Design a one-way deterministic finite state machine to show that the set of all strings of A, B, (and) in which the parentheses are matched, but which contain no nestings of parentheses is a regular language.

Exercise 8 Design a one-way deterministic finite state machine to show that the set of all strings of A, B, (and) in which the parentheses are matched, but which contain no nestings of parentheses of depth greater than three is a regular language. For example A(((AB))A)() should be accepted but ((((A)B))) shouldn't. How could your machine be simplified if it were to accept only those strings which contain no nestings of depth greater than two?

The point of this exercise is to give you a feeling for why the set of matched strings is not a regular language. The greater the depth of nesting to be tested, the more states you would need for your machine. When you have finished building your finite state machine, go back and open Paren Check to remind yourself how this Turing machine manages to recognize this language.

Exercise 9 Build a one-way, deterministic finite state machine that accepts the same language accepted by the machine you built in Exercise 3, page 82.

Exercise 10 Build the two-way, deterministic finite state machine with only one state, an accepting state, with two arcs. Label one of these arcs (A, ⇒) and the other (B, ⇐). What language is accepted by this little machine? Design a one-way finite state machine that accepts the same language.

8.2 Turing machines

Exercise 11 (Food chain simulation) Design a Turing machine that will simulate action in the food chain. Imagine that a B always "eats" any A that it's facing and moves forward (that is, right), taking its place. (When the B moves forward, everyone behind him should follow.) Then, after all the B's have eaten, a C always eats any B that it's facing and moves forward. For example, the sequence CACBBAAA would turn into CAC. Your machine should transform any tape with a string of A's, B's and C's into one that represents the result of everyone's having eaten.

Exercise 12 (Copier) In building complex machines, it is often important to be able to copy a string of symbols. Build a machine which acts as follows. It starts on the left of a string α of A's and B's and ends on the left of the string $\alpha\alpha$, that is, it ends with two copies of α, with no space between them. You may use auxiliary symbols.

Exercise 13 Insert the copier into the holes in the Iterater schematic machine described in Chapter 6, page 74, and see if it works as expected.

Exercise 14 Redo Exercise 12 without using any symbols other than A and B. Compare the complexity of the two machines in terms of the number of states and arcs. Compare the time-space count of the two when they each copy the string

BBABAB

to get the string

BBABABBBABAB.

Exercise 15 (A^nB^n) Build a Turing machine that begins at the left of a (possibly empty) string of A's and B's, and writes GOOD if the string is of the form A^nB^n (i.e., has zero or more A's followed by the very same number of B's), but writes BAD otherwise. This is something that cannot be done by a finite state machine, as suggested in Exercise 6.

Exercise 16 (Cancelation) Exercise 4 asked you to combine our three prepackaged machines into one that represented multiplication in any Abelian group with the two elements A and B. Now suppose that in addition we assume $A = B^{-1}$; that is, that AB is the multiplicative identity, so that you cancel any A with a B next to it. Design a machine that will accept any string of A's, B's, ('s, and)'s, and do as before, except this time canceling occurrences of AB. This exercise raises a general question about the computability of operations in a group. For more on this important topic, the reader is referred to Davis's chapter in [Barwise].

Exercise 17 (Coding) In Chapter 3, page 56, we described a way to encrypt the alphabet A, B, − with the alphabet *, −. Build a Turing machine to carry out this encryption, one that takes an arbitrary string of A's, B's and −'s and replaces it with the encrypted version. Assume that the input string is enclosed in parentheses, and likewise have the output delimited with parentheses. (What problems would come up if we didn't assume this?) Your machine should start and stop on the left parenthesis.

Exercise 18 (Decoding) Build a decoding machine for the system of encryption used in Exercise 17.

Exercise 19 Put together the machines designed in Exercises 17 and 18 so that they take a string, code it, and then decode it back into the original string.

Exercise 20 (Coding using the default alphabet) In Exercise 17, we really cheated, since we used auxiliary parentheses in our alphabet. Design a coding machine that will allow us to do without any auxiliary symbols. To do this, you need to assume that the input string does not contain two or more blanks in a row. In other words, A−−B is not a legitimate input, though A−B−B is. Your machine should start on the first symbol of the input string (which could be a blank) and should halt on the first square of the resulting code (which again may be blank).

8.3 Numerical computation

Given any Turing machine M on the alphabet $*$ and blank, we will say that M computes the (possibly partial) function f_M^2 of two numerical arguments, where this function is determined as follows. Given numbers j, k, form a tape with $j + 1$ $*$'s, a blank, then $k + 1$ $*$'s. We use one extra $*$ so that 0 is represented by something definite, a single $*$, rather than by the blank, which is used to separate the arguments. Position the read head at the leftmost $*$ and run the machine. If the machine eventually halts with the read head at the left end of a string of $m + 1$ $*$'s (on an otherwise blank tape), then we will say that $f_M^2(j, k) = m$. If the machine does not halt, or halts in some other configuration, then $f_M^2(j, k)$ is not defined. (If f_M^2 is defined for all arguments, then it is said to be a *total* function; if not it is *partial*.)

We can generalize this for functions of other than two arguments in the natural way. Any given machine M can be said to compute a function f_M^n of n arguments: Given natural numbers $k_1, \ldots, k_n$, form a tape with $k_1 + 1$ $*$'s, a blank, then $k_2 + 1$ $*$'s, a blank, and so on. Position the read head at the leftmost $*$ and run the machine. If the machine halts, and in doing so has the read head at the left end of a string of $m + 1$ $*$'s (on an otherwise blank tape), then $f_M^n(k_1, \ldots, k_n) = m$. If the machine does not halt, or halts in some other configuration, then $f_M^n(k_1, \ldots, k_n)$ is not defined.

The conventions we've just described for determining the function computed by a Turing machine are called the *standard input-output conventions*. The leftmost nonblank symbol on the input tape is said to be the *standard starting position*. Since Turing machines simply manipulate marks on a tape, it's clear that we have to adopt some such conventions in order to talk about the machines computing functions. Other conventions are possible, of course.

If an n-ary function f from natural numbers to natural numbers is of the form f_M^n for some Turing machine M, then f is said to be *Turing computable*. *Turing's Thesis* is the claim that the Turing computable functions are exactly those functions on natural numbers that are intuitively computable.

Exercise 21 (Zero function) Show that the function $f(k) = 0$ is Turing computable by building a Turing machine to compute it. In other words, build a machine that takes as input a single string of $*$'s and, when started in the standard starting position, eventually halts scanning a single $*$ on an otherwise blank tape.

Exercise 22 (Successor) Show that the function $f(k) = k + 1$ is Turing computable by building a machine to compute it. How few states can you get away with?

Exercise 23 (Addition) Build a Turing machine to compute the addition function $f(k, m) = k + m$.

Exercise 24 (Collapsing variables) Build a schematic Turing machine which shows that if $f(m, n)$ is Turing computable, so is the function $g(n) = f(n, n)$. (We can think of this as showing that we can "collapse" any binary computable function to a unary computable function. For example, if you plug in a multiplying machine, you would end up with a squaring machine.)

Exercise 25 (Doubling) Apply the schematic machine from Exercise 24 to your addition machine from Exercise 23 to get a Turing machine that computes the function $double(n) = n + n$. Save this machine for future reference.

Exercise 26 (Partial subtraction) Since we are dealing only with natural numbers, the function $f(n, m) =$

$n - m$ is a partial function: it is only defined for $m \leq n$. Show that this partial function is Turing computable. (If $m < n$ then your machine should never halt, or else halt in some nonstandard configuration.) You may use auxiliary symbols.

Exercise 27 (Partial subtraction with no auxiliary symbols) Redo Exercise 26 without using any auxiliary symbols. This can be done using only nine states. (See [Davis].) See how few states you can get by with.

Exercise 28 (Proper subtraction) Proper subtraction, $\dot{-}$, is defined so that if $n \geq m$ then $n \dot{-} m = n - m$, but otherwise $n \dot{-} m = 0$. Modify the machine for subtraction to show that proper subtraction is Turing computable.

Exercise 29 (Identity (projection) functions) Define, for $n \geq i$, the function $id_i^n(x_1, \ldots, x_n) = x_i$. Show that these are all Turing computable. Do this by building the machine for id_2^4 and explaining how you would modify it to get the other cases.

Exercise 30 (Multiplication) Build a Turing machine to compute the multiplication function.

Exercise 31 (Doubling, revisited) In Exercise 25 we built a machine to compute the function $double(n) = n + n$. However, we also know that $double(n) = 2n$. (a) Use the machine from Exercise 30 to build another machine for computing the function $double$. The machine should begin by putting three $*$'s, and a blank to the left of the input string representing n, and then running your multiplier. (b) Compare the time efficiency of your two doublers as follows. On a single graph, plot the number of steps it takes each of them to compute $double(n)$ for $n = 0, 1, \ldots, 10$. Explain why one machine is so much more efficient than the other.

Exercise 32 (Transfer machine) In proving things

about Turing computable functions, it is important to be able to show that you can always rearrange the order of the arguments in any way you want without affecting computability. Design a machine which will start with a sequence of five strings $\alpha_1\ \alpha_2\ \alpha_3\ \alpha_4\ \alpha_5$ of *'s, with each successive pair separated by one blank, and rearrange them in the order $\alpha_3\ \alpha_4\ \alpha_5\ \alpha_1\ \alpha_2$. (For a solution, see [Davis].)

Exercise 33 (5-tuple machines) In Chapter 6, page 76, we discussed a different formalization of the notion of a Turing machine, one where a machine could both print and move in one step. Such a machine would need to have arcs labeled by triples, not pairs, and so need five items to specify one state transition, not four. For example the 5-tuple, $3\ *\ -\ \Leftarrow\ 4$, would represent the command: in state 3 if you see a *, print a blank, move left, and go to state 4. You can always simulate any such machine with a 4-tuple machine.

(a) Come up with a general technique for transforming any 5-tuple machine into a 4-tuple machine. (Assume that the machines use only the default alphabet.)

(b) Apply your technique to the 5-tuple machine given by:

$$0 * * \Rightarrow 0$$
$$0 - * \Rightarrow 1$$
$$1 * * \Rightarrow 1$$
$$1 - - \Leftarrow 2$$
$$2 * - \Leftarrow 3$$
$$3 * - \Leftarrow 4$$
$$4 * * \Leftarrow 4$$
$$4 - - \Rightarrow 5$$

This machine, which computes a familiar numerical function f of two arguments, is borrowed from Enderton's chapter of [Barwise], which is based on the 5-tuple approach. Run your simulation on a few input tapes and see if you can figure out what f is. Compare the machine you have just built with the machine you built earlier to compute the same function. Which uses the most states? Compare the time-space count of the two when used to compute $f(4,4)$ and $f(8,8)$.

8.4 Recursive functions

If we start with the zero function (from Exercise 21), the successor function (from Exercise 22) and the identity functions (from Exercise 29), we generate what is known as the class of *recursive functions* on the natural numbers by means of three function building rules—rules which get us new functions from old functions. The rules are known as *composition*, *primitive recursion* and *minimalization*. For a precise definition, see [Boolos & Jeffrey], [Davis], or [Kleene]. The following exercises form the basis of a proof that every recursive function is Turing computable. The converse is also true, so in fact the Turing computable functions are exactly the recursive functions. This is part of the evidence for Turing's Thesis.

Exercise 34 We're going to outline the proof that all recursive functions are Turing computable. To prove this, it turns out that it is easier to show a bit more, that all recursive functions are computable by Turing machines that never move left of the square on which they are started, and also end up with the read head on that same initial square. Let's call such machines "R-machines."

Go back and show that the zero function, the successor function and the identity functions are all computable by R-machines. You may use an auxiliary symbol.

Exercise 35 (Unary composition) Show that if $f(x)$ and $g(x)$ are computable by R-machines, then so is $h(x) = f(g(x))$. Do this by building a schematic machine (see page 74) with two submachine holes, so that if you are given any R-machines to compute f and g, you could simply turn them into submachines and paste them into your machine at the submachine nodes and get an R-machine that computes h.

Exercise 36 (Binary composition) Unary composition is just one case of a more general process of composing functions. The case where we compose by substituting two functions is more difficult. Show that if $f(x, y)$, and $g_1(x)$ and $g_2(x)$ are all computable by R-machines, then so is $h(x) = f(g_1(x), g_2(x))$. Again, do this by building a schematic machine, this time with three submachines for holes. (This is a bit tricky, but it can be done.)

Exercise 37 (Primitive recursion) A very important method for defining functions is that of *primitive recursion*, where the value of the function can be computed using "earlier" values of the function. For example, given functions $f(x)$ and $g(x, z)$, we can define a new function $h(x, y)$ by

$$h(x, 0) = f(x)$$

$$h(x, y + 1) = g(x, h(x, y)).$$

Show that if f and g are computable by R-machines, so is h. Do this by building a schematic machine PrimRec for h into which you could paste submachines for the functions f and g.

Exercise 38 Define a function h by primitive recursion as follows:

$$h(x, 0) = 0,$$

$$h(x, y + 1) = x + h(x, y).$$

1. Prove that $h(x, y) = x \times y$. [*Hint:* Prove this by induction on y.]
2. Design an R-machine that computes the addition function.

3. Plug your zero R-machine and your addition R-machine into PrimRec to get another function which computes the multiplication function. Be sure to test it.

4. Do a graphical analysis of the number of steps it takes each of your multipliers to compute $x \times x$ for $x = 0, 1, \ldots, 10$.

5. If you are using this in a class, compare your results with those of other students. One of the lessons that should emerge is that what seems a small inefficiency in the design of a template machine can explode into a massive inefficiency in some applications of it.

Exercise 39 (**The Ackermann function**) Functions that can be defined from the basic functions using only composition and primitive recursion are often call *primitive recursive*. There was a period of time when it was thought that the primitive recursive functions might exhaust the computable functions. However, the German logician Wilhelm Ackermann came up with a binary function $A(x, y)$ which is clearly computable, but which "grows faster" than any binary primitive recursive function. That is, for any primitive recursive function $f(x, y)$ there is an integer n such that $A(x, y) > f(x, y)$ for all $x, y > n$.

Open the state diagram Ackermann function. This Turing machine computes A. Run it on the tape $* - *$ to compute $A(0, 0)$. What is the answer? Go ahead and determine $A(1, 1)$, $A(2, 2)$, and $A(3, 3)$. If you have nothing else for your computer to do for quite a while, you might try running this machine to compute $A(4, 4)$. (Remember, you can always pause your machine during a computation and reset it. You will need to since the answer is $2^{4,294,967,296} - 3$.)

As we have said, the Ackermann function is not primitive recursive. However, once you have the operation of minimalization, introduced in the next exercise, you can define the Ackermann function from the basic functions. We will not ask you to do this.

Exercise 40 (Minimalization) Another method for defining functions is that of *minimalization*. For example, given a function $f(x, y)$, define $g(x)$ to be the least y such that (1) $f(x, y) = 0$ and (2) for all $z < y$, $f(x, z)$ is defined and not equal to 0. This definition is often written $g(x) = \mu y(f(x, y) = 0)$, where μ (the Greek letter mu) stands for "minimum." Build a schematic machine that would allow you to compute any h defined in this manner, one where you can simply paste an R-machine in for f.

Project 41 We have just sketched half of the proof that the Turing computable functions are exactly the recursive functions, specifically, the half that shows that recursive functions are all Turing computable. As we mentioned earlier, the fact that both of these characterizations specify the same set of functions is often seen as evidence for Turing's Thesis, the claim that Turing computable functions are exactly those that are intuitively computable.

It turns out, though, that there are many other equivalent characterizations of this same class of functions, some involving other types of machines (for example, register machines or abacus machines [Boolos & Jeffrey]) and others involving alternative ways of defining functions (for example, Church's λ-calculus or Herbrand-Gödel equation schemata [Kleene]). The equivalence of all of these different approaches adds up to even more striking evidence for Turing's Thesis. A valuable project would be to learn about one of these other attempts to characterize the class of computable functions, and then to convince yourself that all functions of the given class are Turing computable. In most cases this would involve designing schematic machines of the sort sketched in the above exercises.

8.5 The busy beaver function

Consider, for any natural number n, the problem of building a Turing machine with exactly n states, which uses only the default alphabet, and which behaves like this: Started on an empty tape, it works away until it eventually halts with the read head on the leftmost * of a sequence of consecutive *'s (on an otherwise blank tape). That's easy enough. But now imagine a contest to build such an n-state machine which prints the largest possible number of *'s when started on an empty tape. For any fixed n there must be such a largest number, since there are only a finite number of n-state Turing machines. Call this number $bb(n)$. Of all machines with n states, those that write $bb(n)$ *'s and then halt in the right place are called n-state *busy beavers* (hence the name of the function bb). The function bb is one of the simplest examples of a function that is not Turing computable. (And so, if Turing's Thesis is correct, it is not computable.) For a proof of this, see either [Boolos & Jeffrey] or Enderton's chapter of [Barwise].

Exercise 42 (Busy Beaver No. 2) Show that $bb(2) \geq 2$ by building a machine with two states which, when started on an empty tape, writes two *'s and halts on the leftmost of these. In fact, $bb(2) = 2$, and so your machine is a 2-state busy beaver. But how would you prove this?

Exercise 43 (Busy Beaver No. 7) Build a machine with seven states that shows that $bb(7) > 7$. Do you think your machine is a 7 state busy beaver? Are you sure?

Exercise 44 (Busy Beaver No. 6) Show that $bb(6) > 6$.

Exercise 45 (Busy Beaver n+1) Show that $bb(n + 1) > bb(n)$, for any n.

Exercise 46 (Busy Beaver No. 6, revisited) See if you can show that $bb(6) > 10$. After you have done your best, compare your result with the machine in Greg's Challenge, on your Turing's World disk. If your machine bested Greg's, please send us a copy.

8.6   The Halting Problem

Turing's original paper [Turing] contained all the basic results about Turing machines and has had untold consequences, both theoretical and practical. (See [Herken].) For example, it contained a proof of the undecidability of the Halting Problem, one of the most important and basic results in computability theory. Also, the idea of a programmable computer is implicit in this paper, with Turing's notion of a "universal" Turing machine, one that can simulate any other Turing machine.

The basic idea behind both of these results is that any Turing machine M can itself be represented by a symbolic expression α_M in some predetermined way. For example, we could represent it as a sequence of 4-tuples or, since this sequence can be coded as a natural number, M could even be represented by a string α_M of $*$'s in the default alphabet. Given any such symbolic representation of Turing machines, a number of interesting problems arise.

Define the numerical function $halt(m, n)$ as follows: $halt(m, n) = 1$ if the Turing machine M_m coded by m, when run on a string of $n + 1$ $*$'s, eventually halts. Otherwise $h(m, n) = 0$.

Is $halt$ Turing computable? That is, is there a Turing machine H which starts on the left of a string α_M–β and works away, eventually ending at the left of a sequence of one or two $*$'s, two if the machine M would halt when run on a tape which contained β, one if M would not halt on this tape? Turing showed that there is no such Turing machine. That is, the function *halt* is not Turing computable.

The proof that *halt* is not Turing computable is by means of a so-called "diagonal argument." Start with any old Turing machine H and let $h(m, n)$ be the binary function it computes with the standard input-output conventions. Think of it as someone's candidate for a machine to compute *halt*. We need to show that $h \neq halt$. Since *halt* is total, we need only consider the case where h is total, since otherwise we are already done. The idea is to define a new "diagonal machine" D_H, and then get a contradiction by running D_H on an appropriate tape. If we can make sure that for every m, D_H halts on a string of $m + 1$ *'s just in case $h(m, m) = 0$, then we will know that $h \neq halt$. Why? Well otherwise, just assume that m is the number which codes the machine D_H. There will be such a number, but both of the assumptions $h(m, m) = halt(m, m) = 0$ and $h(m, m) = halt(m, m) \neq 0$ lead to contradictions.

Exercise 47 (Diagonal machine) We have given you the state diagram of a schematic machine of D on your Turing's World disk, called Diagonalizer. This machine has a submachine node, node 7, which is left blank, so that you can fill it in with any possible candidate for a halting machine H. Take any machine H which computes a total two-place function h under the standard input-output conventions. (You have built many such machines in doing the earlier problems.) Copy H and paste it into node 7 of D. Run the resulting version D_H of D on several different strings. Verify that D_H halts on a string of $m + 1$ *'s if and only if $h(m, m) = 0.$[1]

Project 48 (Diagonal machine for busy beaver)

---

[1]The schematic machine D is a slight modification of a machine used in the proof in [Boolos & Jeffrey].

Design a schematic machine, analogous to that from Exercise 47, which will demonstrate that bb is not Turing computable. (See [Boolos & Jeffrey] if you need help on this project.)

8.7 Universal Turing machines

As we mentioned earlier, Turing's original paper introduced the important notion of a universal Turing machine. Consider some fixed way of representing Turing machines by strings of symbols, and consider the problem of taking a Turing machine representation α_M, and another input sequence β, and asking for the result of running the machine M on β. A universal machine U is one that would start with any such α_M and β as input and carry out the simulation. You can think of the machine U as a general purpose programmable computer, and think of α_M as a program for doing what the special purpose computer M would do alone. Turing showed that there are indeed universal Turing machines.

The project of designing a universal Turing machine is an excellent one to carry out. Many of our students have come up with very elegant solutions and in doing so have gained a great insight into Turing machines and the programmability of modern digital computers.

The first step in designing a universal Turing machine is to settle on a coding scheme to use to represent Turing machines. We will only code machines that use the default alphabet, that do not contain submachines, and whose start state is state 0. For simplicity, though, our universal machine will use both auxiliary symbols and submachines. There are various other tradeoffs that can be made in choosing a coding scheme. Some make it easy to code machines and some make the operation of the machine more elegant. One of the smallest universal machines has been designed by Marvin Minsky and is reproduced in [Haugland], but the coding Minsky used is far from transparent.

By contrast, we have chosen a fairly transparent coding scheme, one that makes it easy to code machines and to use these codes. Here's a recipe for getting the code of a machine.

1. Obtain the 4-tuple description of the machine.

2. Represent each 4-tuple by a sequence of the form

   S...SIAN...N

   where we use S...S to code up the state number, N...N to code up the successor state, and where I and A are single symbols representing the input symbol and action, respectively. For example, the 4-tuple

   $$0 * \Rightarrow 1$$

   will be represented as the sequence

   S*RNN

   As you can see, we are using $n + 1$ S's and N's to represent state n. We use L and R to represent $\Leftarrow$ and $\Rightarrow$, and use * and the blank to represent themselves.

3. String the representations of all of the 4-tuples together, with no extra spaces, arranged in increasing order of state numbers. If there are two 4-tuples starting with the same state number, put the one with input symbol * before the one whose input is the blank.

4. Finally, wrap the result in a pair of parentheses.

Let's illustrate this coding with the Pacer, whose 4-tuple description is as follows:

$$0 * \Rightarrow 0$$
$$0 - \Leftarrow 1$$
$$1 * \Leftarrow 1$$
$$1 - \Rightarrow 0$$

The code for the pacer would be the following string:

(S*RNS–LNNSS*LNNSS–RN)

Exercise 49 Open the tape TM1 Code. Build the machine that is coded on this tape. What function is computed by this machine?

Project 50 Build a universal Turing machine that uses the above coding scheme to represent the simulated machine. Your universal machine should start on an input tape that consists of a code for a machine, followed by an input string α located somewhere to the right of the machine code. Initially, the read head should be located on the same symbol of α as the computation to be simulated. After running on this input, the universal machine should halt if and only if the simulated machine halts on input α. If the simulated machine halts with string ω appearing on the final tape, then the universal machine should halt with ω appearing to the right of the machine code, and the read head should end up on the corresponding symbol.

For example, when you run your universal machine on the tape TM1 Code with input, it should eventually end up with the tape shown in TM1 Code with output.

One of the things you will have to worry about in designing your universal machine, is what happens when your simulated machine moves left and bumps into its own code. In this eventuality, you will have to shift the machine code to the left to give your simulated machine room to move. As a result, your output tape may contain some extra blank space between the machine code and ω.

In doing this project, plan the whole thing carefully before you start building any of the pieces. It will save you a lot of work. Think about how to package various frequently used routines into submachines. Annotate these submachines clearly to tell your instructor (and remind yourself) what role things are supposed to be playing.

Exercise 51 Open the tape TM1 Code with input and run your universal machine on it. The result should look like the tape TM1 Code with output.

Exercise 52 Open the tape TM2 Code with input and run your universal machine on it. If your universal machine is working correctly, it should end up with a string of five consecutive *'s somewhere to the right of the machine code, with the read head on the rightmost *. The machine simulated here will give your shift left routine a workout.

Exercise 53 Open the tape TM3 Code with input. Here you will find the code of a machine that computes the identity function Id_1^2 (see Exercise 29), along with the input string **−**. Run your universal Turing machine on this tape.

For further discussion of universal Turing machines, see [Turing], [Davis], or [Haugland].

8.8   Computability and logic

There are many important topics connecting computability theory with topics in propositional and predicate logic. Turing's World makes it easy to investigate these topics by allowing alphabets to contain the basic symbols of logic: $\land, \lor, \lnot, \rightarrow, \forall$, and $\exists$. In this section (and later) we present a few exercises and projects that touch on these topics. The reader unfamiliar with propositional logic should consult [Barwise & Etchemendy] or some other standard textbook on logic in doing these problems.

Exercise 54 (Propositional wffs) Let's restrict attention to the symbols $\land, \lor$, and $\lnot$. If we are given some set L of basic propositional letters, the set of *well-formed formulas* (wffs), is defined inductively as follows:

1. Every letter in L is a wff.
2. If φ and ψ are wffs, so are $(\varphi \wedge \psi)$ and $(\varphi \vee \psi)$.
3. If φ is a wff, so is $\neg\varphi$
4. Nothing is a wff except in virtue of (1)–(3).

Design a Turing machine that recognizes the set of wffs. For simplicity, let's assume that the basic propositional letters are P, Q, and R. Create a custom alphabet that contains the symbols $P, Q, R, (,), \wedge, \vee$, and $\neg$, plus some auxiliary symbols for use in your computations. Your Turing machine should operate as follows. It starts on the left end of a string of the symbols just mentioned and does one of two things. If the string is a wff, the machine halts with the tape unchanged and the read head again at the left end of the string. If the string is not a wff, the machine erases the string and replaces it with the string NOT A WFF.

Project 55 Generalize the machine just constructed so that it recognizes the wffs built out of propositional letters $P, P', P'', \ldots$. Use the symbol $*$ for the prime symbol.

Exercise 56 Build a Turing machine corresponding to the truth table for the connective $\wedge$. Your machine should operate as follows. Starting on a wff built out of the propositional letters P, Q, R, T, F, it should make one pass through the wff, replacing any wff which is a conjunction of T and F with the value assigned to it by the standard truth table for $\wedge$. In particular, it should replace each substring of the form $(T \wedge T)$ by T, those of the form $(T \wedge F)$ by F, and so on. It should remove any blank spaces caused by these replacements.

Exercise 57 Build a Turing machine corresponding to the truth table for the connective ∨. Your machine should operate like the one designed in Exercise 56, except using the truth table for ∨.

Exercise 58 Build a Turing machine corresponding to the truth table for the operator ¬. Your machine should operate like the one designed in Exercise 56, except using the truth table for ¬.

The following project is not hard conceptually, but it does take some time to get the details right. It will be used in a number of the remaining exercises and is well worth the effort.

Project 59 (Evaluation of wffs) Design a machine that checks to see if a given wff built from P, Q, and R is true under a given truth assignment. Your machine should act as follows. It starts on a string consisting of a wff built out of P, Q, and R, followed by a blank, followed by a string of T's and F's of length three. This string is to represent a truth assignment. For example, if the string is TFT, it represents a situation where the atomic sentence P is true, Q is false, and R is true. The standard truth table interpretation of the logical symbols gives an associated truth value to the whole wff. Your machine should replace the string of three truth values by this derived truth value, ending with the read head at the left end of the original wff. For example, if the input is (P ∧ (Q ∨ R)) TFT then the output should be (P ∧ (Q ∨ R)) T, since the wff is true if P and R are true and Q is false. [Hint: Use the machines designed in Exercises 56, 57, and 58.]

Exercise 60 (Satisfiability) A wff is *satisfiable* if there is a truth assignment under which it is true. Build a Turing machine that determines whether an arbitrary wff built from P, Q, and R is satisfiable. Your machine should start at the left of a wff and end by printing

SAT or NOT SAT one space to the right of the wff. Use the machine designed in Project 59, by systematically searching through the possible truth value assignments. [Note: there are only eight possible truth assignments. However, in general, if we have n basic propositional letters, there are 2^n possible truth assignments. This causes the length of time needed for this approach to satisfiability to be exponential in the number of propositional letters. It is an open question whether one can find a deterministic method which is not exponential, but polynomial—that is, where there is some fixed number k (independent of n) so that the computation always takes less that n^k steps. Most experts doubt there is. However, if we allow nondeterministic machines, then there is such a number. We will indicate the reason in Exercise 69.]

Exercise 61 (Validity) A wff is *valid* if for each assignment of truth values to atomic letters, its associated truth value is T. Build a machine that determines whether or not a wff is valid. Your machine should start at the left of a wff and end by printing VALID or NOT VALID one space to the right of the wff. [Hint: Use the machine designed in Exercise 60, using the fact that a wff is valid if and only if its negation is not satisfiable.]

The following project is also worthwhile, but unlike Project 59, it will not be used in later problems.

Project 62 (Conjunctive normal form) A wff is said to be a *literal* if it is either a propositional letter or ¬x for some propositional letter x. A wff is in *conjunctive normal form* (CNF) if it is a conjunction of disjunctions of literals. Every wff is logically equivalent to one in CNF. (See [Barwise & Etchemendy], Chapter 3, or some other standard logic text.) Design a Turing machine which starts on the left of a wff and ends by writing an equivalent CNF wff one square to the right of the original wff. [Hint: design submachines that perform operations like pushing a negation through a conjunction ($\neg(\varphi \wedge \psi)$ becomes ($\neg\varphi \vee \neg\psi$)).]

8.9   Nondeterministic machines

An important theorem in computability theory shows that any nondeterministic Turing machine can be mimicked by a deterministic Turing machine. The same holds true of nondeterministic finite state machines. So in one sense, the nondeterministic machines do not give you any additional computing power. But they sometimes make it much easier to tackle a problem. Moreover, it has been conjectured that nondeterministic machines allow you to compute certain problems much more efficiently. (See Exercise 69.)

Exercise 63 Open the state diagram Mystery Language and run it on some strings of A's and B's. What language does this nondeterministic finite state machine accept? Build a deterministic finite state machine that accepts the same language.

Exercise 64 (Regular languages, revisited) Assume that the alphabet contains just A and B.

1. Design a deterministic finite state machine that accepts the set of strings in which every third symbol from the left is a B.

2. Design a nondeterministic finite state machine that accepts the set of strings in which every third symbol *from the right* is a B. (In other words, strings whose reverse is in the previous language.)

3. Design a *deterministic* machine that accepts the same language as your nondeterministic machine from (2).

Exercise 65 Open the state diagram Stuck on B's. Notice that this is a two-way, deterministic finite state machine. Figure out what language this machine accepts by running it on various strings of A's and B's. Then build a one-way, nondeterministic finite state machine that accepts the same language.

Exercise 66 (Maze runner) Build a nondeterministic Turing machine that, when started at any position in a 6 × 6 maze, finds the shortest path to a goal G. Represent the maze using *'s for the walls and G for the goal, as in the following:

```
*  *  *  *  *  *
*              *
*     *  *  *  *
*        *  G  *
*              *
*  *  *  *  *  *
```

Then, to fit this on the tape, put the rows one after another, with ='s between them. Thus, the above maze should appear on the tape as follows:

$$* * * * * = * - - - - * = * - * * * * = * - - * G * =$$
$$* - - - - * = * * * * * *$$

Design your machine so that when you start the read head on a blank inside a 6 × 6 maze, it runs until it finds the G, leaving a record (say a trail of crumbs) along the shortest path from where it started to the goal. Your machine should only move horizontally and vertically, not diagonally, and it should always halt, even on mazes where the goal is inaccessible from the starting position. If it finds a G, it should immediately go into a halt state, to stop the computation. You may use auxiliary symbols.

You can try out your machine on the saved tapes Maze Tape 1, Maze Tape 2, and Maze Tape 3. In one of these, there is no path to the goal.

Exercise 67 Build a deterministic Turing machine that produces the same result as the three state nondeterministic machine you built in Exercise 7. (If you did not do that exercise earlier, do it now.) Use as few states as possible.

Exercise 68 Design a nondeterministic Turing machine with alphabet consisting of A and B such that the machine has a successful computation if and only if the input string can be written in the form $\alpha\beta\gamma$ where each of these is a string, and where γ is just β in reverse. α is allowed to be empty, but β and γ are not. [Hint: Imagine walking down the tape with a host of assistants. At each square you tell a different assistant to take responsibility for determining whether the string ends in the reverse of a string starting at the current position.]

Exercise 69 (Nondeterministic test for satisfiability) Build a nondeterministic version of the Turing machine required in Exercise 60. Your machine should operate as follows. It starts at the left of a wff. After working its way right, it splits into two computations, one corresponding to the case where P is assigned T, the other where P is assigned F. Each of these then splits corresponding to the cases where Q is T or F. Finally, each of these splits into two cases, depending on the assignment of truth value to R. This gives us a total of eight cases. On each of these, your machine should run the machine created in Project 59.

This problem is related to a famous open problem in computer science, known as the P=NP problem. Notice that, in general, a successful computation showing that a wff is satisfiable will be much shorter than that found in Exercise 60, since we are doing all eight in parallel, not one after the other. A simple generalization of your machine shows that the satisfiability problem for propositional wffs is in the class NP, those problems that can be solved in polynomial time using nondeterministic machines. Is it in the class P? That is, can you construct a deterministic machine that solves it in polynomial time? Of course the general problem is for wffs in an arbitrary number of propositional letters, not just three. For more on this problem, see, for example, [Hopcroft & Ullman].

Appendix: Macintosh Terminology

The following is a brief glossary of some Macintosh terminology used in this book.

Active window At any time, only one window on screen is "active." This is indicated by the lines in its title bar.

Application A computer program designed for doing a particular kind of thing. For example, Turing's World is an application, as is the Finder, from which you start up Turing's World. You are probably also familiar with word processing applications, if you have ever written a paper on the computer.

Choose To choose an item from a menu, pull down the menu and, with the button still depressed, move the mouse cursor until the item is highlighted. Then release the mouse button.

Click To click on an item appearing on the screen, move the mouse until the cursor is on top of the item. Then click the button on the mouse.

Clipboard A temporary storage location, used to store text or other information for transferring to another location.

Close box The close box is the small white box in the upper left corner of the window (at the left end of the window's title bar). Clicking in it closes the window.

Command key The command key is the one with the cloverleaf on it. Newer keyboards also have an apple on the command key. To type (say) Command-A, type A while holding down the command key. In other words, use it like you use the shift key when typing upper-case letters.

Deleting files To delete files, you must be at the desktop (see below), and have the disk's contents window open (this is the window showing the Turing and Machines and Tapes folder icons). You will first need to find the icon for the file you want to delete. If it is in a folder, you will have to open the folder by double-clicking on it. When you find the file, click on it and drag it to the trash can icon. When the trash can turns black, release the button. If you change your mind about deleting the file, retrieve it (before you do anything else) by double-clicking on the trash can and moving the file back to the contents window. If the machine you are working on is running System 7, the trash can is only emptied when you choose **Empty Trash...** from the **Special** menu. You must do this if you want more space on your disk.

Desktop The desktop is the screen that appears when you are in the Finder. It is the screen with the trash can icon in the lower right corner.

Dialog box A rectangular box that appears on the screen to request or provide information.

Dragging To drag an item from one point on the screen to another, move the mouse until the cursor is on top of the item. Click on the item and, with the button still depressed, move the mouse until the item is where you want it. Then release the button.

Double-click To double-click on an item, move the mouse until the cursor is on top of the item. Then click the button on the mouse twice in quick succession.

File Turing's World has two kinds of files: state diagram files and tape files. These are created in the appropriate windows and can be saved on the disk.

Finder The finder is the program that starts up when you first turn on your Macintosh. It displays the trash can icon in the lower right corner and the disk icon(s) in the upper right.

Folder To create a folder, you must be in the Finder. Choose **New Folder** from the Finder's **File** menu, and an empty folder will appear in the contents window. Change the folder's name to the desired name.

Highlight Highlighting is often used to indicate a selected item. Visually, this is indicated by changing the usual colors of the item. For example, black text on a white background becomes white text on a black background when it is selected. This is known as highlighting.

Icon Icons are the graphical representations used by the Macintosh for things like programs, files, disks, and the like.

Initialized disk When you buy a blank disk, it has to be initialized before it can be used. You do this from the Finder (see above). Put the new disk in the spare drive (if your Macintosh has two), and it will ask you if you want to initialize it. Click the appropriate button. It will ask you to name the disk; choose some name other than "Turing's World," to prevent confusion. If you have only one drive, choose **Eject** from the **File** menu and put the new disk in the drive. Then follow the same procedure.

Insertion point The vertical line that appears in the block of text when you are writing annotations. It indicates where characters will be inserted when you enter them.

Menu A menu is a list of commands or functions that appears when you click on its title. The menu titles are at the top of the screen, to the right of the apple.

Pull down To pull down a menu, click on the menu title at the top of the screen. Do not release the button until the desired item is chosen. If you decide not to choose an item, simply move the mouse cursor away from the menu.

Scroll Scrolling a window allows you to see portions that are not presently visible. To scroll a window, either click on the up or down arrows in the scroll bar to the right of the window, or drag the scroll box, which appears between the arrows, up or down.

Scroll bar A grey bar with arrows at the top and bottom, and a small square (the scroll box) that moves up and down the bar. The scroll bar is used to move the contents of a window up and down.

Select To select an item, you simply click on it. The item will become highlighted. (In many cases, some item, say the first in a list, will already be highlighted. If that is the one you want to select, then you don't have to do anything.)

Sizing box This is the little box in the lower right corner of the Sentence window. containing two overlapping squares. To change the size of one of the windows, click in this box and drag.

Title bar The title bar is the bar across the top of a window, usually containing the title of the window or the name of the file displayed in the window.

Window Windows are the rectangular areas on the screen in which all activity must take place.

Index

CSLI Publications

CSLI Publications are distributed
world-wide by Cambridge University Press
unless otherwise noted.

Lecture Notes

A Manual of Intensional Logic. van
Benthem, 2nd edition. No. 1.
0-937073-29-6 (paper), 0-937073-30-X

*Lectures on Contemporary Syntactic
Theories.* Sells. No. 3. 0-937073-14-8
(paper), 0-937073-13-X

The Semantics of Destructive Lisp.
Mason. No. 5. 0-937073-06-7 (paper),
0-937073-05-9

An Essay on Facts. Olson. No. 6.
0-937073-08-3 (paper), 0-937073-05-9

Logics of Time and Computation.
Goldblatt, 2nd edition. No. 7.
0-937073-94-6 (paper), 0-937073-93-8

*Word Order and Constituent Structure in
German.* Uszkoreit. No. 8.
0-937073-10-5 (paper), 0-937073-09-1

*Color and Color Perception: A Study in
Anthropocentric Realism.* Hilbert.
No. 9. 0-937073-16-4 (paper),
0-937073-15-6

Prolog and Natural-Language Analysis.
Pereira and Shieber. No. 10.
0-937073-18-0 (paper), 0-937073-17-2

*Working Papers in Grammatical Theory
and Discourse Structure: Interactions
of Morphology, Syntax, and Discourse.*
Iida, Wechsler, and Zec (Eds.). No. 11.
0-937073-04-0 (paper), 0-937073-25-3

*Natural Language Processing in the 1980s:
A Bibliography.* Gazdar, Franz,
Osborne, and Evans. No. 12.
0-937073-28-8 (paper), 0-937073-26-1

Information-Based Syntax and Semantics.
Pollard and Sag. No. 13.
0-937073-24-5 (paper), 0-937073-23-7

Non-Well-Founded Sets. Aczel. No. 14.
0-937073-22-9 (paper), 0-937073-21-0

Partiality, Truth and Persistence.
Langholm. No. 15. 0-937073-34-2
(paper), 0-937073-35-0

*Attribute-Value Logic and the Theory of
Grammar.* Johnson. No. 16.
0-937073-36-9 (paper), 0-937073-37-7

The Situation in Logic. Barwise. No. 17.
0-937073-32-6 (paper), 0-937073-33-4

The Linguistics of Punctuation. Nunberg.
No. 18. 0-937073-46-6 (paper),
0-937073-47-4

*Anaphora and Quantification in Situation
Semantics.* Gawron and Peters.
No. 19. 0-937073-48-4 (paper),
0-937073-49-0

*Propositional Attitudes: The Role of
Content in Logic, Language, and
Mind.* Anderson and Owens. No. 20.
0-937073-50-4 (paper), 0-937073-51-2

Literature and Cognition. Hobbs. No. 21.
0-937073-52-0 (paper), 0-937073-53-9

*Situation Theory and Its Applications,
Vol. 1.* Cooper, Mukai, and Perry
(Eds.). No. 22. 0-937073-54-7
(paper), 0-937073-55-5

*The Language of First-Order Logic
(including the Macintosh program,
Tarski's World 4.0).* Barwise and
Etchemendy, 3rd Edition. No. 23.
0-937073-99-7 (paper)

Lexical Matters. Sag and Szabolcsi (Eds.).
No. 24. 0-937073-66-0 (paper),
0-937073-65-2

Tarski's World: Macintosh Version 4.0.
Barwise and Etchemendy. No. 25.
1-881526-27-5 (paper)

*Situation Theory and Its Applications,
Vol. 2.* Barwise, Gawron, Plotkin, and
Tutiya (Eds.). No. 26. 0-937073-70-9
(paper), 0-937073-71-7

Literate Programming. Knuth. No. 27.
0-937073-80-6 (paper), 0-937073-81-4

*Normalization, Cut-Elimination and the
Theory of Proofs.* Ungar. No. 28.
0-937073-82-2 (paper), 0-937073-83-0

Lectures on Linear Logic. Troelstra.
No. 29. 0-937073-77-6 (paper),
0-937073-78-4

A Short Introduction to Modal Logic.
Mints. No. 30. 0-937073-75-X
(paper), 0-937073-76-8

Linguistic Individuals. Ojeda. No. 31.
0-937073-84-9 (paper), 0-937073-85-7

*Computational Models of American
Speech.* Withgott and Chen. No. 32.
0-937073-98-9 (paper), 0-937073-97-0

Verbmobil: A Translation System for Face-to-Face Dialog. Kay, Gawron, and Norvig. No. 33. 0-937073-95-4 (paper), 0-937073-96-2

The Language of First-Order Logic (including the Windows program, Tarski's World 4.0). Barwise and Etchemendy, 3rd edition. No. 34. 0-937073-90-3 (paper)

Turing's World. Barwise and Etchemendy. No. 35. 1-881526-10-0 (paper)

The Syntax of Anaphoric Binding. Dalrymple. No. 36. 1-881526-06-2 (paper), 1-881526-07-0

Situation Theory and Its Applications, Vol. 3. Aczel, Israel, Katagiri, and Peters (Eds.). No. 37. 1-881526-08-9 (paper), 1-881526-09-7

Theoretical Aspects of Bantu Grammar. Mchombo (Ed.). No. 38. 0-937073-72-5 (paper), 0-937073-73-3

Logic and Representation. Moore. No. 39. 1-881526-15-1 (paper), 1-881526-16-X

Meanings of Words and Contextual Determination of Interpretation. Kay. No. 40. 1-881526-17-8 (paper), 1-881526-18-6

Language and Learning for Robots. Crangle and Suppes. No. 41. 1-881526-19-4 (paper), 1-881526-20-8

Hyperproof. Barwise and Etchemendy. No. 42. 1-881526-11-9 (paper)

Mathematics of Modality. Goldblatt. No. 43. 1-881526-23-2 (paper), 1-881526-24-0

Feature Logics, Infinitary Descriptions, and Grammar. Keller. No. 44. 1-881526-25-9 (paper), 1-881526-26-7

Tarski's World: Windows Version 4.0. Barwise and Etchemendy. No. 45. 1-881526-28-3 (paper)

German in Head-Driven Phrase Structure Grammar. Pollard, Nerbonne, and Netter. No. 46. 1-881526-29-1 (paper), 1-881526-30-5

Formal Issues in Lexical-Functional Grammar. Dalrymple and Zaenen. No. 47. 1-881526-36-4 (paper), 1-881526-37-2

Dynamics, Polarity, and Quantification. Kanazawa and Piñón. No. 48. 1-881526-41-0 (paper), 1-881526-42-9

Theoretical Perspectives on Word Order in South Asian Languages. Butt, King, and Ramchand. No. 50. 1-881526-49-6 (paper), 1-881526-50-X

Perspectives in Phonology. Cole and Kisseberth. No. 51. 1-881526-54-2 (paper)

Linguistics and Computation. Cole, Green, and Morgan. No. 52. 1-881526-81-X (paper)

Modal Logic and Process Algebra: A Bisimulation Approach. Ponse, de Rijke, and Venema. No. 53. 1-881526-96-8 (paper)

Dissertations in Linguistics Series

Phrase Structure and Grammatical Relations in Tagalog. Kroeger. 0-937073-86-5 (paper), 0-937073-87-3

Theoretical Aspects of Kashaya Phonology and Morphology. Buckley. 1-881526-02-X (paper), 1-881526-03-8

Argument Structure in Hindi. Mohanan. 1-881526-43-7 (paper), 1-881526-44-5

The Syntax of Subjects. Tateishi. 1-881526-45-3 (paper), 1-881526-46-1

Theory of Projection in Syntax. Fukui. 1-881526-35-6 (paper), 1-881526-34-8

On the Placement and Morphology of Clitics. Halpern. 1-881526-60-7 (paper), 1-881526-61-5

The Structure of Complex Predicates in Urdu. Butt. 1-881526-59-3 (paper), 1-881562-58-5

Configuring Topic and Focus in Russia. King. 1-881526-63-1 (paper), 1-881562-62-3

The Semantic Basis of Argument Structure. Wechsler. 1-881526-68-2 (paper), 1-881562-69-0

Stricture in Feature Geometry. Padgett. 1-881526-66-6 (paper), 1-881562-67-4

Possessive Descriptions. Barker. 1-881526-72-0 (paper), 1-881562-73-9

Studies in Logic, Language and Information

Logic Colloquium '92. Csirmaz, Gabbay, and de Rijke (Eds.). 1-881526-98-4 (paper), 1-881526-97-6

Meaning and Partiality. Muskens.
1-881526-79-8 (paper), 1-881526-80-1

Logic and Visual Information. Hammer.
1-881526-99-2 (paper), 1-881526-87-9

Other CSLI Titles Distributed by Cambridge University Press

The Proceedings of the Twenty-Fourth Annual Child Language Research Forum. Clark (Ed.). 1-881526-05-4 (paper), 1-881526-04-6

The Proceedings of the Twenty-Fifth Annual Child Language Research Forum. Clark (Ed.). 1-881526-31-3 (paper), 1-881526-33-X

The Proceedings of the Twenty-Sixth Annual Child Language Research Forum. Clark (Ed.). 1-881526-31-3 (paper), 1-881526-33-X

Japanese/Korean Linguistics. Hoji (Ed.). 0-937073-57-1 (paper), 0-937073-56-3

Japanese/Korean Linguistics, Vol. 2. Clancy (Ed.). 1-881526-13-5 (paper), 1-881526-14-3

Japanese/Korean Linguistics, Vol. 3. Choi (Ed.). 1-881526-21-6 (paper), 1-881526-22-4

Japanese/Korean Linguistics, Vol. 4. Akatsuka (Ed.). 1-881526-64-X (paper), 1-881526-65-8

The Proceedings of the Fourth West Coast Conference on Formal Linguistics (WCCFL 4). 0-937073-43-1 (paper)

The Proceedings of the Fifth West Coast Conference on Formal Linguistics (WCCFL 5). 0-937073-42-3 (paper)

The Proceedings of the Sixth West Coast Conference on Formal Linguistics (WCCFL 6). 0-937073-31-8 (paper)

The Proceedings of the Seventh West Coast Conference on Formal Linguistics (WCCFL 7). 0-937073-40-7 (paper)

The Proceedings of the Eighth West Coast Conference on Formal Linguistics (WCCFL 8). 0-937073-45-8 (paper)

The Proceedings of the Ninth West Coast Conference on Formal Linguistics (WCCFL 9). 0-937073-64-4 (paper)

The Proceedings of the Tenth West Coast Conference on Formal Linguistics (WCCFL 10). 0-937073-79-2 (paper)

The Proceedings of the Eleventh West Coast Conference on Formal Linguistics (WCCFL 11). Mead (Ed.). 1-881526-12-7 (paper),

The Proceedings of the Twelfth West Coast Conference on Formal Linguistics (WCCFL 12). Duncan, Farkas, Spaelti (Eds.). 1-881526-33-X (paper),

The Proceedings of the Thirteenth West Coast Conference on Formal Linguistics (WCCFL 13). Aranovich, Byrne, Preuss, Senturia (Eds.). 1-881526-76-3 (paper),

European Review of Philosophy: Philosophy of Mind. Soldati (Ed.). 1-881526-38-0 (paper), 1-881526-53-4

Experiencer Subjects in South Asian Languages. Verma and Mohanan (Eds.). 0-937073-60-1 (paper), 0-937073-61-X

Grammatical Relations: A Cross-Theoretical Perspective. Dziwirek, Farrell, Bikandi (Eds.). 0-937073-63-6 (paper), 0-937073-62-8

Theoretical Issues in Korean Linguistics. Kim-Renaud (Ed.). 1-881526-51-8 (paper), 1-881526-52-6

Agreement in Natural Language: Approaches, Theories, Descriptions. Barlow and Ferguson (Eds.). 0-937073-02-4

Papers from the Second International Workshop on Japanese Syntax. Poser (Ed.). 0-937073-38-5 (paper), 0-937073-39-3

Ordering Titles from Cambridge University Press

Titles distributed by Cambridge University Press may be ordered directly from the distributor at 110 Midland Avenue, Port Chester, NY 10573-4930 (USA), or by phone: 914-937-9600, 1-800-872-7423 (US and Canada), 95-800-010-0200 (Mexico). You may also order by fax at 914-937-4712.

Overseas Orders

Cambridge University Press has offices worldwide which serve the international community.

Australia: Cambridge University Press, 120 Stamford Road, Oakleigh, Victoria 31266, Australia. phone: (613) 563-1517. fax: 613 563 1517.

UK, Europe, Asia, Africa, South America: Cambridge University Press, Publishing Division, The Edinburgh Building, Shaftesbury Road, Cambridge CB2 2RU, UK.

Inquiries: (phone) 44 1223 312393 (fax) 44 1223 315052

Orders: (phone) 44 1223 325970 (fax) 44 1223 325959

CSLI Titles Distributed by The University of Chicago Press

The Phonology-Syntax Connection. Inkelas and Zec. 0-226-38100-5 (paper), 0-226-38101-3

On What We Know We Don't Know. Bromberger. 0-226-07540-0 (paper), 0-226-07539-7

Arenas of Language Use. Clark. 0-226-10782-5 (paper), 0-226-10781-7

Head-Driven Phrase Structure Grammar. Pollard and Sag. 0-226-67447-9 (paper)

Titles distributed by The University of Chicago Press may be ordered directly from UCP. Phone 1-800-621-2736. Fax (800) 621-8471.

Titles distributed by The University of Chicago Press may be ordered directly from UCP. Phone 1-800-621-2736. Fax (800) 621-8471.

CSLI Titles Distributed by CSLI Publications

Hausar Yau Da Kullum: Intermediate and Advanced Lessons in Hausa Language and Culture. Leben, Zaria, Maikafi, and Yalwa. 0-937073-68-7 (paper)

Hausar Yau Da Kullum Workbook. Leben, Zaria, Maikafi, and Yalwa. 0-93703-69-5 (paper)

Reports

Coordination and How to Distinguish Categories Sag, Gazdar, Wasow, and Weisler 84-3

Belief and Incompleteness Konolige 84-4

Equality, Types, Modules and Generics for Logic Programming Goguen and Meseguer 84-5

Lessons from Bolzano van Benthem 84-6

Self-propagating Search: A Unified Theory of Memory Kanerva 84-7

Reflection and Semantics in LISP Smith 84-8

The Implementation of Procedurally Reflective Languages des Rivières and Smith 84-9

Parameterized Programming Goguen 84-10

Completeness of Many-Sorted Equational Logic Goguen and Meseguer 84-15

Moving the Semantic Fulcrum Winograd 84-17

On the Mathematical Properties of Linguistic Theories Perrault 84-18

A Simple and Efficient Implementation of Higher-order Functions in LISP Georgeff and Bodnar 84-19

On the Axiomatization of "if-then-else" Guessarian and Meseguer 85-20

The Situation in Logic–II: Conditionals and Conditional Information Barwise 84-21

Principles of OBJ2 Futatsugi, Goguen, Jouannaud, and Meseguer 85-22

Querying Logical Databases Vardi 85-23

Computationally Relevant Properties of Natural Languages and Their Grammar Gazdar and Pullum 85-24

An Internal Semantics for Modal Logic: Preliminary Report Fagin and Vardi 85-25

The Situation in Logic–III: Situations, Sets and the Axiom of Foundation Barwise 85-26

Ordering Titles Distributed by CSLI

Titles distributed by CSLI may be ordered directly from CSLI Publications, Ventura Hall, Stanford, CA 94305-4115. Orders can also be placed by FAX (415)723-0758 or e-mail (pubs@csli.stanford.edu).

All orders must be prepaid by check or Visa or MasterCard (include card name, number, and expiration date). California residents add 8.25% sales tax. For shipping and handling, add $2.50 for first book and $0.75 for each additional book; $1.75 for first report and $0.25 for each additional report.

For overseas shipping, add $4.50 for first book and $2.25 for each additional book; $2.25 for first report and $0.75 for each additional report. All payments must be made in U.S. currency.

READ THIS INFORMATION BEFORE OPENING THE POUCH CONTAINING THE DISKETTE

Please note that this book and the double-side, double-density diskette accompanying it may not be returned after the pouch containing the diskette has been unsealed. The diskette, Turing's World 3.0, is protected by copyright and may not be duplicated, except for purposes of archival back up. All inquiries concerning the diskette should be directed to the CSLI Publications, Ventura Hall, Stanford University, Stanford, CA 94305; phone: (415)723–1712 or (415)723–1839; or via e-mail at pubs@roslin.stanford.edu. The program has been duplicated on the finest quality diskettes, verified before shipment. If a diskette is damaged, please send it to the above address for replacement. Please note also that the distributor of this book and program, Cambridge University Press, is not equipped to handle inquiries about this package, nor to replace defective diskettes.

A NOTE TO BOOKSTORES

The publisher recommends that bookstores not sell used copies of this text-software package. Bookstores selling used copies must be prepared to warrant that (1) the original owner retains no copies of the software and (2) the diskette contains all of the exercise files in their original form and contains no solution files created by the initial owner.